W9-CZG-015

THE ENVIRONMENT

THE
ENVIRONMENT

A National Mission
for the Seventies

by

The Editors of FORTUNE

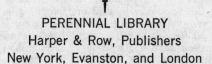

PERENNIAL LIBRARY
Harper & Row, Publishers
New York, Evanston, and London

CONTENTS

The contents of this book originally appeared in the pages of FORTUNE magazine. The authors acknowledge with gratitude the editorial guidance of Robert Lubar, Richard Armstrong, Walter Guzzardi, Thomas O'Hanlon, and the contributions of the following members of FORTUNE's research staff: Lucie Adam, Evelyn Benjamin, Jo Degener, Carol Higgins, Helen Howard, Margaret Kennedy, Jeanne Krause, Ann Tyler, Marilyn Wellemeyer, Bernice Zimmerman. Charts were designed by the art department.

Introduction:

Reconciling Progress with the Quality of Life

Within the recent past an immense transformation has occurred in public concern about the environment. One scientist describes the shift with a metaphor drawn from geology: "First, islands of anxiety about specific environmental ills—like the redwoods, the rivers, or the slums—rose from a sea of apathy; when they rose further, land appeared between them; we became aware that all these separate environmental issues were connected, all part of a single challenge to our civilization." Now that environmental anxieties have coalesced, they will be a permanent part of the American awareness, part of the set of beliefs, values, and goals within which U.S. business operates.

The new awareness brings a danger of its own, stepping up the urgency of our situation. Unless we demonstrate, quite soon, that we can improve our environmental record, U.S. society will become paralyzed with shame and self-doubt.

Already we hear voices—and not merely from noisy rebels among the young—exploiting our environmental anxieties as part of an indictment against the basic characteristics and trends of Western civilization. The idea of material progress is especially deplored. "What this country needs," says a professor, "is Buddhist economics." He explains that we ought to welcome a yearly decline in the gross national product, leading us back toward the economy of the Asiatic village, say in Tibet, without factories or much transport.

Nobody really thinks that Minnesota will swap Mustangs for yaks. Nevertheless, the sentimental rhetoric of the simple-lifers takes on, as our environmental anxiety rises, a dangerous capacity to distract us from the challenge that we actually face.

We know that isolated societies with very low levels of technology do not greatly damage their natural environments. We also know that our high-technology society is handling our environment in a way that will be lethal for us. What we don't know—and had better make haste to test—is whether a high-technology society can achieve a safe, durable, and improving relationship with its environment. This—and not a return to the pre-technological womb—is the only possibility worth investigating.

The broad outlines of what such an effort entails are sketched in Chapter XIII "How to Think About the Environment." Looked at one by one, many of our present depredations seem relatively easy to correct. But when we put the horrors in a row—the drab and clumsy cities, the billboards, the scum-choked lakes, the noise, the poisoned air and water, the clogged highways, the mountainous and reeking dumps—their cumulative effect drives us toward the conclusion that some single deep-seated flaw in modern society is responsible for all of these.

This is true. The defect, worse luck, is intimately bound up with the virtues of modern society, with science and democracy and individuality and diversity and prosperity. Because our strength is derived from the fragmented mode of our knowledge and our action, we are relatively helpless when we try to deal intelligently with such unities as a city, an estuary's ecology, or "the quality of life." Since we will not retreat to older forms of unity, we have to seek it at a new level by such drastic innovations as restructuring the market, the government, and the university.

The most important news in this book is that environmental reform is going to be harder to achieve than many of its advocates suggest. The difficulties may be more political than economic. Although many specific environmental policy decisions will require careful analysis, an over-all benefit-cost calculation of a decent environment would be very difficult to make. We have no idea how large a proportion of our present production serves only to compensate for the disutilities and diseconomies created by other parts of our production. During the next fifty years we will replace, anyway, nearly all of our man-made environment. Will it cost more—or less—to do it right? Will it cost more—or less—to deal carefully with nature? The cost of a decent environment might be enormous by conventional accounting standards while its true social cost would be zero.

What the U.S. needs to work upon first are the political and other arrangements by which the environment might be protected and improved. Almost every article in this issue questions, explicitly or implicitly, the adequacy of our present decision-making processes. The host of environmental problems centering upon the automobile, examined in "Cars and Cities on a Collision Course," Chapter V, nearly all raise questions of how new forms of public regulation might shift market directions so people could get more of what they really want and less of what they don't want. As Henry Ford II has pointed out, government standards are often necessary to create a market, an arena of innovation and choice, as in the case of emission-control devices on cars.

The eloquent statements addressed to the reader by President Nixon and Senator Muskie are remarkable for their similarity. Improvement of the environment is not a partisan issue. Nor is it a class issue. Some militant academics who believe that class conflict must always be the mainspring of politics and of history argue that "the environment" is a middle-class cause, intended to divert government funds that should be distributed to the poor.

"Hungry people don't care about the environment," says one class warrior.

That's the kind of thinking that got us into this mess. In a high-technology society, the single-minded pursuit of any goal—even such a worthy one as feeding the hungry —is almost always bound to produce undesirable side effects on the environment. Unless we learn to watch for and prevent the side effects, all of our past and future efforts toward material progress and social justice will be futile.

To reverse environmental deterioration will be one of the main goals of the next generation, involving all the major functions of society. It will not be enough to cope with each environmental atrocity as it reaches the point of clear and present danger to life and health. This book sets forth the proposition that the U.S. must start inventing political, economic, and intellectual processes that will give us, as a society and as individuals, more real choice about how we live.

A Statement from President Nixon

From our nation's very beginnings, the generous hand of nature has played a major role in our development. As the historian David Potter has written, we have been a "people of plenty," and the abundance of our country has been a cardinal factor in shaping our national character. The breadth and variety and beauty of our land, the richness of our mines and soil and forests and water, the favorable nature of our climate—all of these natural factors have provided a setting in which the optimism, the ingenuity and drive of the American people thrive and grow and are rewarded.

The combination of natural and human resources, in turn, has given substance to the promise of equality and opportunity for all. A generous land has produced a generous and confident people, a people who have dreamed great dreams of personal, national, and human destiny and who have seen many of these dreams come true. Alexis de Tocqueville summarized the point this way in his study of America in the mid-nineteenth century:

"Their ancestors gave them the love of equality and of freedom; but God Himself gave them the means of remaining equal and free by placing them upon a boundless continent."

The abundance that has helped to make America great, however, has too often worked to make us careless. Many Americans have taken our resources for granted and have relied, without reflection, on their continued plenty. Many have failed to imagine a time when the air and water would lose their sparkling quality, when each man's living space would itself begin to shrink under the pressure of growing population, when the beauty of life on our continent would be marred by offensive sights and sounds and odors. And because these problems were often unanticipated, too little effort was made to prepare for them or to forestall them. Even as we are heirs to the products of our forefathers' genius, so our generation has also inherited the results of past carelessness.

Almost two centuries ago, the founding fathers defined the American dream as "life, liberty and the pursuit of happiness." America's first century was devoted to establishing and preserving our life as a nation. Our second century—one which will soon be ended—can be thought of as a time when liberty was extended to more and more people. The major concern of our third century, I believe, will be with the pursuit of happiness. In conducting that pursuit, we must remember that happiness is not measured in quantitative but rather in qualitative terms. It is not achieved merely by piling up objects or by adding years to our life; it is accomplished by enriching each moment of our experience, by bringing new life to our years.

In recent years, many Americans—and particularly young Americans—have become increasingly aware of the part the natural environment plays in determining the quality of their lives. Perhaps no single goal will be more important in our future efforts to pursue the public happiness than that of improving our environment. This goal, I believe, is one that will help define a new spirit for the Seventies, a new expression of our idealism, a new challenge that will test our ingenuity. I am confident that this goal can be met, but only if we give it the same priority Theodore Roosevelt did when he described the conservation and proper use of natural resources as "the fundamental problem which underlies almost every other problem of our national life."

If we are to materially improve our environment in the months and years ahead, then *all* of our people must join in that effort. Strong government action will be required—at the federal, state, and local level. Private citizens and voluntary groups must join in the crusade. So must businesses and industries, labor and farm organizations, educational and scientific institutions, and every part of our society. It is important, too, that the quality of our environment be seen not only as a national but also as an international concern. Now that we have seen our planet as it appears from outer space, all men of all nations can appreciate it more clearly than ever before as their common home, small and round and one, with a thin and precious atmosphere on which we all depend and with no artificial boundaries to divide our energies or keep us apart.

Man has applied a great deal of his energy in the past to exploring his planet. Now we must make a similar commitment of effort to restoring that planet. Our scientific capacity has grown so much that we are able to leave the earth; yet our glimpse of its beauty from the barren moon has only reminded us of how much we must love earth's qualities. The unexpected consequences of our technology have often worked to damage our environment; now we must turn that same technology to the work of its restoration and preservation.

If we can do these things, then the coming decade will be not only the beginning of our third century as a nation, but also the time of the renewal of America's sense of infinite promise.

. . . and from Senator Muskie

Environment has become a popular catchword. Conservationists, students, businessmen, housewives, physicians, scientists, and politicians use it and exalt it. What used to be the battle against the polluters is now the fight to improve the quality of our environment.

The change in words underscores a change in perceptions about the nature of our problems. The degree of concern illustrates the quantum jump in public understanding of the real threat pollution poses to man's survival and his health. The challenge to public and private leaders is to respond with more than pious agreement and vague promises.

That response will not be easy. The environment is comprehensive and complex. It is the air we breathe, the water we drink, the noise we hear, the buildings, trees, flowers, oceans, lakes, rivers, and open spaces we view and through which we move, and the vehicles which move us. Our every action affects that environment, and through our ability to extend the application of energy and to manipulate the physical world we have magnified our effects on it.

We have just scratched the surface of understanding the nature of the changes we have made in our environment, and as we seek to correct the abuses of the past we find even bigger threats looming over us. While we build treatment plants to eliminate the gross discharges of sewage and other organic wastes, we are aware of the fact that lakes are "dying" as a result of man's cumulative assaults on waterways—"dying" as a result of complex processes we understand only in part. While we install electrostatic precipitators to reduce the discharge of particulates from factories and power plants, we are aware of the threat of noise and atmospheric changes from the SST.

We are confronted with the terrible prospect that the American dream of the good life may turn out to be a nightmare. Our efforts to improve our lives may have created hazards from which there is no escape. From this time forward we must devote as much energy and ingenuity to the elimination of man-made hazards to man as we have to the expansion of his ability to harness energy and materials to his desires.

The guidelines for such a course are not as clear as we might like. Many of the hazards to man and his environment are not obvious, and their impact may not be known for many years. Our experience with cigarettes and radiation hazards provides two examples of the long-term, low-level effects of contaminants on individuals. There is, moreover, the difference between being non-sick and being healthy. The latter requires an attractive, pleasant, and stimulating man-made environment, coupled with a natural environment which is in a state of dynamic balance.

All of this adds up to the need for an environmental policy which is designed to correct the abuses of the past, to eliminate such abuses in the future, to reduce unnecessary risks to man and other forms of life, and to improve the quality of our design and development of communities, industrial units, transportation systems, and recreational areas. Such a policy must be carried out in the context of an increasing population which, because of the leisure and affluence available to it, will make greater demands on resources.

We are not well equipped to undertake such a policy today. Federal, state, and local jurisdictions overlap in some cases and leave gaps in others. Appropriate relations have not been established between land and water use patterns, tax systems, and urban designs. Economic activities which provide short-term gains at the expense of long-term public interest are not susceptible to deliberate decisions as to their desirability.

Congress has been engaged in an effort to develop systems for environmental improvement. We have a base on which to build in the Air and Water Quality Acts, the Solid Waste Act, and the Environmental Quality Act. The immediate success of those programs depends on the resources we are willing to commit for their implementation. The Water Quality Act, for example, requires an appropriation of $1.24 billion in fiscal 1971. Beyond this we must upgrade the capacity of the federal government to organize itself and to ensure consistency in its attack. I have recommended the creation of an independent environmental protection agency to marshal the resources necessary to combat the interlocking assaults on our air, water, and land resources. Such an agency would reflect the national commitment we need if we are to avoid ecological disaster.

The establishment of such an agency must be backed by a commitment of resources to eliminate the discharge of municipal and industrial wastes into our public waterways, to achieve drastic reduction of air pollution emissions, to prevent the distribution of materials and products which threaten life, and to ensure the reconstruction and development of our metropolitan areas.

That commitment cannot be achieved by public officials alone. It will require the concerted effort and dedication of men and women from every walk of life, an effort and dedication I hope this book will stimulate.

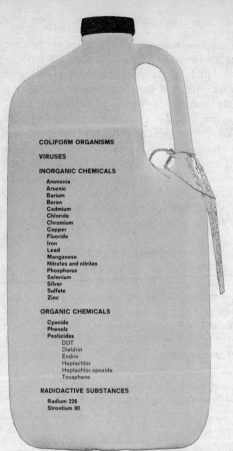

COLIFORM ORGANISMS

VIRUSES

INORGANIC CHEMICALS

Ammonia
Arsenic
Barium
Boron
Cadmium
Chloride
Chromium
Copper
Fluoride
Iron
Lead
Manganese
Nitrates and nitrites
Phosphorus
Selenium
Silver
Sulfate
Zinc

ORGANIC CHEMICALS

Cyanide
Phenols
Pesticides
  DDT
  Dieldrin
  Endrin
  Heptachlor
  Heptachlor epoxide
  Toxaphene

RADIOACTIVE SUBSTANCES

Radium 226
Strontium 90

Beautiful Ohio was an often heard song of yesteryear. But the Ohio River no longer lives up to that kind of image. The water itself is not very pretty when subjected to laboratory analysis: It contains hundreds, perhaps thousands, of simple and complex pollutants. Some of them are listed on the sampling bottle, filled with water dipped up near Cincinnati. Coliform bacteria and related organisms come from sewage. Viruses include the forms that cause polio and hepatitis. Some of the substances listed can be toxic even in small quantities. About 100 million Americans get their drinking water from rivers and lakes that may contain such pollutants. The water usually goes through some kind of treatment before people drink it, but the treatment has little effect on most of the pollutants.

I

The Limited War on Water Pollution

by Gene Bylinsky

To judge by the pronouncements from Washington, we can now start looking forward to cleaner rather than ever dirtier rivers. The Administration has declared a "war" on pollution, and Secretary of the Interior Walter J. Hickel says, "We do not intend to lose." Adds Murray Stein, enforcement chief of the Federal Water Pollution Control Administration: "I think we are on the verge of a tremendous cleanup."

The nationwide campaign to clean up ravaged rivers and lakes does seem to be moving a bit. For the first time since the federal government got into financing construction of municipal sewage plants in 1956, Congress has come close to providing the kind of funding it had promised. There are other signs that the war is intensifying. Under the provisions of the Water Quality Act of 1965 and the Clean Water Restoration Act of 1966, federal and state officials are establishing water-quality standards and plans for their implementation, to be carried out eventually through coordinated federal-state action. Timetables for new municipal and industrial treatment facilities are being set, surveillance programs are being planned, and tougher federal enforcement authority is being formulated. Without waiting for these plans to materialize, Interior is talking tough to some municipal and industrial polluters, with the possibility of court action in the background.

Even with all this, however, the water-pollution outlook is far from reassuring. Although the nation has invested about $15 billion since 1952 in the construction of 7,500 municipal sewage-treatment plants, industrial treatment plants, sewers, and related facilities, a surprising 1,400 communities in the U.S., including good-sized cities like Memphis, and hundreds of industrial plants still dump untreated wastes into the waterways. Other cities process their sewage only superficially, and no fewer than 300,000 industrial plants discharge their used water into municipal sewage plants that are not equipped to process many of the complex new pollutants.

Since the volume of pollutants keeps expanding while water supply stays basically the same, more and more intervention will be required just to keep things from getting worse. Within the next fifty years, according to some forecasts, the country's population will double, and the demand for water by cities, industries, and agriculture has tended to grow even faster than the population. These water uses now add up to something like 350 billion gallons a day (BGD), but by 1980, by some estimates, they will amount to 600 BGD. By the year 2000, demand for water is expected to reach 1,000 BGD, considerably exceeding the essentially unchanging supply of dependable fresh water, which is estimated at 650 BGD. More and more, water will have to be reused, and it will cost more and more to retrieve clean water from progressively dirtier waterways.

Just how bad water pollution can get was dramatically illustrated in the summer of 1969 when the oily, chocolate-brown Cuyahoga River in Cleveland burst into flames, nearly destroying two railroad bridges. The Cuyahoga is so laden with industrial and municipal wastes that not even the leeches and sludge worms that thrive in many badly polluted rivers are to be found in its lower reaches. Many other U.S. rivers are becoming more and more like that flammable sewer.

Even without human activity to pollute it, a stream is never absolutely pure, because natural pollution is at work in the form of soil erosion, deposition of leaves and animal wastes, solution of minerals, and so forth. Over a long stretch of time, a lake can die a natural death because of such pollution. The natural process of eutrophication, or enrichment with nutrients, encourages the growth of algae and other plants, slowly turning a lake into a bog. Man's activities enormously speed up the process.

But both lakes and rivers have an impressive ability to purify themselves. Sunlight bleaches out some pollutants. Others settle to the bottom and stay there. Still others are consumed by beneficial bacteria. These bacteria need oxygen, which is therefore vital to self-purification. The oxygen that sustains bacteria as well as fish and other organisms is replenished by natural aeration from the atmosphere and from life processes of aquatic plants.

Trouble starts when demand for dissolved oxygen exceeds the available supply. Large quantities of organic pollutants such as sewage alter the balance. Bacteria feeding upon the pollutants multiply and consume the oxygen. Organic debris accumulates. Anaerobic areas develop, where microorganisms that can live and grow without free oxygen decompose the settled solids. This putrefaction produces foul odors. Species of fish sensitive to oxygen deficiency can no longer survive. Chemical, physical, and biological characteristics of a stream are altered, and its water becomes unusable for many purposes without extensive treatment.

Pollution today is very complex in its composition, and getting more so all the time. In polluted streams and lakes hundreds of different contaminants can be found: bacteria and viruses; pesticides and weed killers; phosphorus from fertilizers, detergents, and municipal sewage; trace amounts of metals; acid from mine drainage; organic and inorganic

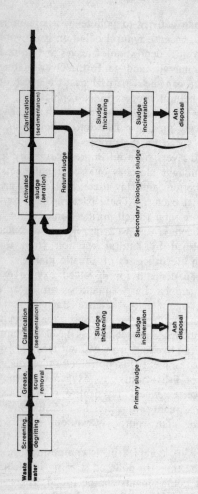

PRIMARY TREATMENT

Waste water → Screening, degritting → Grease, scum removal → Clarification (sedimentation)

Clarification (sedimentation) → Sludge thickening → Sludge incineration → Ash disposal

Primary sludge

SECONDARY

Activated sludge (aeration) → Clarification (sedimentation)

Return sludge

Clarification (sedimentation) → Sludge thickening → Sludge incineration → Ash disposal

Secondary (biological) sludge

How to take out of water some of what people put in. Advanced techniques that remove more subtle pollutants are in use in only a few places in the U.S., and most such plants are still experimental. The operation of one advanced facility, a 7,500,000-gallon-a-day plant at Lake Tahoe in California, is schematically shown above. The waste water passes through three stages, the first two of which generally correspond to the forms of treatment commonly used in the U.S. Metal screens stop large objects such as sticks and rags from entering the plant. The sewage then passes into a grit chamber where sand and small stones settle to the bottom. Next stop is the sedimentation tank, where speed of flow is reduced and suspended particles sink to the bottom, forming sludge. By itself, this primary treatment removes only about 30 percent

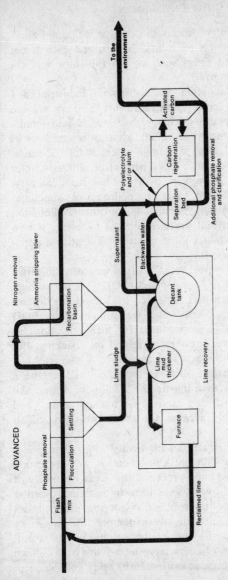

of oxygen-consuming organic matter in sewage. In secondary treatment, most of the remaining organic matter is consumed by bacteria. Aeration speeds up, or "activates," the process. Advanced treatment at Lake Tahoe removes both phosphate and nitrogen, undesirable nutrients that cause proliferation of algae. Phosphate is removed with the help of lime ("flash mix" refers to the rapidity of mixing). Nitrogen, which occurs in sewage mostly as ammonia, is more difficult to eliminate. At Tahoe, the effluent is passed through a stripping tower where ammonia is extracted in a process that involves blowing large amounts of air through the sewage. The effluent then undergoes additional cleansing in passing through separation beds (where chemicals remove more phosphate) and finally through activated carbon. The result is water that's almost good enough to drink.

In the diagram, the following labels appear:

ADVANCED

Phosphate removal
Flash mix
Flocculation
Settling

Nitrogen removal
Ammonia stripping tower
Recarbonation basin

Lime sludge

Lime recovery
Lime mud thickener
Decant tank
Furnace
Reclaimed lime

Supernatant
Backwash water

Polyelectrolyte and/or alum
Separation bed

Additional phosphate removal and clarification

Carbon regeneration
Activated carbon

To the environment

chemicals, many of which are so new that we do not know
their long-term effects on human health; and even traces
of drugs. (Steroid drugs such as the Pill, however, are
neutralized by bacteria.)

A distinction is often made between industrial and
municipal wastes, but it is difficult to sort them out because
many industrial plants discharge their wastes into municipal sewer systems. As a result, what is referred to as
municipal waste is also to a large extent industrial waste.
By one estimate, as much as 40 percent of all waste water
treated by municipal sewage plants comes from industry.
Industry's contribution to water pollution is sometimes
measured in terms of "population equivalent." Pollution
from organic industrial wastes analogous to sewage is now
said by some specialists to be about equivalent to a population of 210 million.

The quality of waste water is often measured in terms
of its biochemical oxygen demand (BOD), or the amount
of dissolved oxygen that is needed by bacteria in decom-
posing the wastes. Waste water with much higher BOD
content than sewage is produced by such operations as
leather tanning, beet-sugar refining, and meatpacking. But
industry also contributes a vast amount of non-degradable,
long-lasting pollutants, such as inorganic and synthetic
organic chemicals that impair the quality of water. All
together, manufacturing activities, transportation, and
agriculture probably account for about two-thirds of all
water degradation.

Industry also produces an increasingly important pollutant of an entirely different kind—heat. Power generation
and some manufacturing processes use great quantities of
water for cooling, and it goes back into streams warmer
than it came out. Power plants disgorging great masses
of hot water can raise the stream temperature by ten or
twenty degrees in the immediate vicinity of the plant.
Warmer water absorbs less oxygen and this slows down
decomposition of organic matter. Fish, being cold-blooded,

cannot regulate their body temperatures, and the additional heat upsets their life cycles; for example, fish eggs may hatch too soon. Some scientists have estimated that by 1980 the U.S. will be producing enough waste water and heat to consume, in dry weather, all the oxygen in all twenty-two river basins in the U.S.

How clean do we want our waterways to be? In answering that question we have to recognize that many of our rivers and lakes serve two conflicting purposes—they are used both as sewers and as sources of drinking water for about 100 million Americans. That's why the new water-quality standards for interstate streams now being set in various states generally rely on criteria established by the Public Health Service for sources of public water supplies. In all, the PHS lists no fewer than fifty-one contaminants or characteristics of water supplies that · should be controlled. Many other substances in the drinking water are not on the list, because they haven't yet been measured or even identified. "The poor water-treatment plant operator really doesn't know what's in the stream—what he is treating," says James H. McDermott, director of the Bureau of Water Hygiene in the PHS. With more than 500 new or modified chemicals coming on the market every year, it isn't easy for the understaffed PHS bureaus to keep track of new pollutants. Identification and detailed analysis of pollutants is just beginning as a systematic task. Only recently has the PHS established its first official committee to evaluate the effects of insecticides on health.

Many water-treatment plants are hopelessly outmoded. They were designed for a simpler, less crowded world. About three-fourths of them do not go beyond disinfecting water with chlorine. That kills bacteria but does practically nothing to remove pesticides, herbicides, or other organic and inorganic chemicals from the water we drink.

A survey by the PHS shows that most waterworks operators lack formal training in treatment processes, disinfection, microbiology, and chemistry. The men are often badly paid. Some of them, in smaller communities, have other full-time jobs and moonlight as water-supply operators. The survey, encompassing eight metropolitan areas from New York City to Riverside, California, plus the State of Vermont, has revealed that in seven areas about 9 percent of the water samples indicated bacterial contamination. Pesticides were found in small concentrations in many samples. In some, trace metals exceeded PHS limits. The level of nitrates, which can be fatal to babies, was too high in some samples. Earlier the PHS found that nearly sixty communities around the country, including some large cities, could be given only "provisional approval" for the quality of their water-supply systems. Charles C. Johnson Jr., administrator of the Consumer Protection and Environmental Health Service in the PHS, concluded that the U.S. is "rapidly approaching a crisis stage with regard to drinking water" and is courting "serious health hazards."

Clearly, there will have to be enormous improvement in either the treatment of water we drink or the treatment of water we discard (if not both). The second approach would have the great advantage of making our waterways better for swimming and fishing and more aesthetically enjoyable. And it is more rational anyway not to put poisons in the water in the first place. The most sensible way to keep our drinking water safe is to have industry, agriculture, and municipalities stop polluting water with known and potentially hazardous substances. Some of this could be accomplished by changing manufacturing processes and recycling waste water inside plants. The wastes can sometimes be retrieved at a profit.

A great deal of industrial and municipal waste water now undergoes some form of treatment. So-called primary treatment is merely mechanical. Large floating objects such as sticks are removed by a screen. The sewage then passes through settling chambers where filth settles to become raw sludge. Primary treatment removes about one-third of gross pollutants. About 30 percent of Americans served by sewers live in communities that provide only this much treatment.

Another 62 percent live in communities that carry treatment a step beyond, subjecting the effluent from primary processing to secondary processing. In this age of exact science, secondary treatment looks very old-fashioned. The effluent flows, or is pumped, onto a "trickling filter," a bed of rocks three to ten feet deep. Bacteria normally occurring in sewage cover the rocks, multiply, and consume most of the organic matter in the waste water. A somewhat more modern version is the activated sludge process, in which sewage from primary settling tanks is pumped to an aeration tank. Here, in a speeded-up imitation of what a stream does naturally, the sewage is mixed with air and sludge saturated with bacteria. It is allowed to remain for a few hours while decomposition takes place. Properly executed secondary treatment will reduce degradable organic waste by 90 percent. Afterward, chlorine is sometimes added to the water to kill up to 99 percent of disease germs.

Secondary treatment in 90 percent of U.S. municipalities within the next five years and its equivalent in most industrial plants is a principal objective of the current war on pollution. The cost will be high: an estimated $10 billion in public funds for municipal treatment plants and sewers and about $3.3 billion of industry's own funds for facilities to treat wastes at industrial plants.

But today that kind of treatment isn't good enough. Widespread use of secondary treatment will cut the amount of gross sewage in the waterways, but will do little to reduce the subtler, more complex pollutants. The effluents will still contain dissolved organic and inorganic contaminants. Among the substances that pass largely unaffected through bacterial treatment are salts, certain dyes, acids, persistent insecticides and herbicides, and many other harmful pollutants.

Technical "tunnel vision," or lack of thinking about all the possible consequences of a process, has often been the curse of twentieth-century science and technology. Today's sewage plants generally do not remove phosphorus and nitrogen from waste water, but turn the organic forms of these nutrients into mineral forms that are *more* usable by algae and other plants. As one scientist has noted, overgrowths of algae and other aquatic plants then rot to "recreate the same problem of oxygen-consuming organic matter that the sewage plant was designed to control in the first place." The multibillion-dollar program to treat waste water in the same old way, he says, is "sheer insanity."

Yet the U.S. has little choice. Most of the advanced treatment techniques are either still experimental or too costly to be introduced widely. To wait for those promising new methods while doing nothing in the meantime could result in a major pollution calamity.

The pollutants that secondary treatment fails to cope with will increase in volume as industry and population grow. Phosphates, for instance, come in large amounts from detergents and fertilizers, and from human wastes. Phosphorus has emerged as a major pollutant only in recent years. Nitrogen, the other key nutrient for algal growth, is very difficult to control because certain blue-green algae can fix nitrogen directly from the air. Since phosphorus is more controllable, its removal from effluents is critically important to limiting the growth of algae.

A few years ago, when it looked as if America's streams and lakes were to become highways of white detergent foam, the manufacturers converted the detergent base from alkyl benzene sulphonate to a much more biologically degradable substance, linear alkylate sulphonate. That effectively reduced the amount of foam but did almost nothing to reduce the amount of phosphates in detergents. The mountains of foam have shrunk, but green mats of algae keep on growing. The developers of deter-

gents failed to consider the possible side effects; such lack
of systematic thinking and foresight is precisely what has
led to today's environmental abuses. It might be possible
to substitute nonphosphorus bases in detergent manufacture—and work is in progress along those lines.

There is little prospect of substituting something else
for the phosphate in fertilizer. It's hard to visualize a fertilizer that is a nutrient when applied to land and not a
nutrient when it enters the water. One way to reduce
water pollution from farmlands would be to reduce the
amounts of chemical fertilizers farmers apply to their
fields—it is the excess fertilizer, not absorbed by plants,
that washes into streams or percolates into groundwater.
Through some complex of social and economic arrangements, farmers might be persuaded to use less fertilizer
and more humus. By improving the texture of soils, as
well as providing slowly released nutrients, humus can
reduce the need for commercial fertilizer to keep up crop
yields. The U.S. produces enormous quantities of organic
wastes that could be converted to humus. Such a remedy
for fertilizer pollution, of course, might seem highly undesirable to the fertilizer industry, already burdened with
excess capacity.

Even if phosphorus pollution from fertilizers and detergents were entirely eliminated—an unlikely prospect—
phosphates from domestic and industrial wastes would
still impose a heavy load upon rivers and lakes. As population and industry grow, higher and higher percentages
of the phosphorus will have to be removed from effluents
to keep the algae problem from getting worse. The conventional technology being pushed by the federal water-
pollution war cannot cope with phosphorus, or with many
other pollutants. But there are advanced technologies that
can. Advanced water treatment, sometimes called "tertiary," is generally aimed at removal of all, or almost all,
of the contaminants.

One promising idea under investigation is to dispense

with the not always reliable bacteria that consume sewage in secondary treatment. Toxic industrial wastes have on occasion thrown municipal treatment plants out of kilter for weeks by killing the bacteria. "We've found that we can accomplish the same kind of treatment with a purely physical-chemical process," says a scientist at the Robert A. Taft Water Research Center in Cincinnati.

In this new approach, the raw sewage is clarified with chemicals to remove most suspended organic material, including much of the phosphate. Then comes carbon adsorption. The effluent passes through filter beds of granular activated carbon, similar to that used in charcoal filters for cigarettes. Between clarification and adsorption, 90 percent or more of the phosphate is removed. The carbon can be regenerated in furnaces and reused. Captured organic matter is burned. Carbon adsorption has the great additional advantage of removing from the water organic industrial chemicals that pass unhindered through biological secondary treatment. The chemicals adhere to the carbon as they swirl through its complex structure with millions of pathways and byways.

Other treatment techniques are under study that make water even cleaner, and might possibly be used to turn sewage into potable water. One of these is reverse osmosis, originally developed for demineralization of brackish water. When liquids with different concentrations of, say, mineral salts are separated by a semipermeable membrane, water molecules pass by osmosis, a natural equalizing tendency, from the less concentrated to the more concentrated side to create an equilibrium. In reverse osmosis, strong pressure is exerted on the side with the greater concentration. The pressure reverses the natural flow, forcing molecules of pure water through the membrane, out of the high-salt or high-particle concentration. Reverse osmosis removes ammonia nitrogen, as well as phosphates, most nitrate, and other substances dissolved in water. Unfortunately, the process is not yet applicable to sewage treatment on a large scale because the membranes become fouled with sewage solids. Engineers are hard at work trying to design better membranes.

New techniques are gradually transforming sewage treatment, technically backward and sometimes poorly controlled, into something akin to a modern chemical process. "We are talking about a wedding of sanitary and chemical engineering," says David G. Stephan, who directs research and development at the Federal Water Pollution Control Administration, "using the techniques of the chemical process industry to turn out a product—reusable water—rather than an effluent to throw away." Adds James McDermott of the Public Health Service: "We're going to get to the point where, on the one hand, it's going to cost us an awful lot of money to treat wastes and dump them into the stream. And an awful lot of money to take those wastes when they are going down the stream and make drinking water out of them. We are eventually going to create treatment plants where we take sewage and, instead of dumping it back into the stream, treat it with a view of recycling it immediately—direct reuse. That is the only way we're going to satisfy our water needs, and second, it's going to be cheaper."

Windhoek, the capital of arid South-West Africa, has gained the distinction of becoming the first city in the world to recycle its waste water directly into drinking water. Waste water is taken out of sewers, processed conventionally, oxidized in ponds for about a month, then run through filters and activated-carbon columns, chlorinated, and put back into the water mains. Windhoek's distinction may prove to be dubious, because the full effects of recycled water on health are unknown. There is a potential hazard of viruses (hepatitis, polio, etc.) being concentrated in recycling. For this reason, many health experts feel that renovated sewage should not be accepted as drinking water in the U.S. until its safety can be more reliably demonstrated.

Costs naturally go up as treatment gets more complex. While primary-secondary treatment costs about 12 cents a thousand gallons of waste water, the advanced techniques in use at Lake Tahoe, for instance, bring the cost up to 30 cents. About 7½ cents of the increase is for phosphorus removal. Reverse osmosis at this stage would raise the cost to at least 35 cents a thousand gallons, higher than the average cost of drinking water to metered households in the U.S. Whatever new techniques are accepted, rising costs of pollution control will be a fact of life.

Ironically, these new treatment techniques, such as removal of phosphorus with chemicals, will intensify one of the most pressing operational problems in waste-water treatment—sludge disposal. Sludge, the solid matter removed from domestic or industrial waste water, is a nuisance, highly contaminated unless it's disinfected. The handling and disposal of sludge can eat up to one-half of a treatment plant's operating budget. Some communities incinerate their sludge, contributing to air pollution. "Now in cleaning the water further we are adding chemicals to take out phosphorus and more solids," says Francis M. Middleton, director of the Taft Center. "While we end up with cleaner water, we also end up with even greater quantities of sludge."

Chicago's struggle with its sludge illustrates some of the difficulties and perhaps an effective way of coping with them. With 1,000 tons of sludge a day to dispose of, the metropolitan sanitary district has been stuffing about half of it into deep holes near treatment plants, at a cost of about $60 a ton. The other half is dried and shipped to Florida and elsewhere where it is sold for $12 a ton to citrus growers and companies producing fertilizers—a nonprofit operation. Vinton W. Bacon, general superintendent of the sanitary district, says this state of affairs can't continue. "We're running out of land. Not only that, but the land we're using for disposal is valuable. And even it will be filled within two years."

Bacon is convinced he has an answer that will not only cut costs but also solve disposal problems indefinitely while helping to make marginal lands bloom. Bacon's scheme, tested in pilot projects in Chicago and elsewhere, is to pump liquid sludge through a pipeline to strip mines and marginal farmland about sixty miles southwest of Chicago. "We put the sludge water through tanks where it's digested," Bacon says. "Then it can be used directly without any odor or health dangers. It's the perfect marriage. That land needs our sludge as much as we need the land. Most astounding, even acquiring the required land at current market prices, taking in the cost of a twenty-four-inch, sixty-mile-long pipeline, the pumps, reservoirs, irrigation equipment, and manpower, the cost would still come to only $20 a ton. We could build a pipeline 200 miles long and still not run higher costs than with our present system."

An aspect of water pollution that seems harder to cope with is the overflow of combined sewers during storms. A combined system that unites storm and sanitary sewers into a single network usually has interceptor sewers, with direct outlets to a stream, to protect the treatment plant from flooding during heavy rains. But in diverting excess water from treatment plants, interceptor sewers dump raw sewage into the waterways. Obviously, this partly defeats the purpose of having treatment plants.

So bad are the consequences of sewer overflow that some specialists would prefer to see part of the federal money being channeled into secondary treatment go into correction of the combined-sewer problem instead. But more than 1,300 U.S. communities have combined sewers, and the cost of separating the systems would be huge. The American Public Works Association estimated the cost of total separation at $48 *billion*. The job could be done in an alternative fashion for a still shocking $15 billion, by building holding tanks for the overflow storm water. Still another possibility would be to build separate systems for sewage and to use existing sewers for storm water. The federal war on water pollution discourages construction of combined sewers but strangely includes no money (except for $28 million already awarded for research and development) to remedy the problem of existing combined-sewer systems.

The General Accounting Office recently surveyed federal activities in water-pollution control and found some glaring deficiencies. The G.A.O. prepared its report for Congress and therefore failed to point out that in some of the deficiencies the real culprit was Congress itself. Still largely rural-oriented, Congress originally limited federal grants for construction of waste-treatment facilities to $250,000 per municipality. The dollar ceiling was eventually raised, but was not removed until fiscal 1968. In the preceding twelve years about half of the waste-treatment facilities were built in hamlets with populations of less than 2,500, and 92 percent in towns with populations under 50,000.

In drafting the legislation that provides for new water-quality standards, Congress again showed limited vision, leaving it up to the states to decide many important questions. Each state is free to make its own decisions on pollution-control goals in terms of determining the uses to which a particular stream or lake will be put. Each state is to decide on the stream characteristics that would allow such uses—dissolved oxygen, temperature, etc. Finally, each state is to set up a schedule for corrective measures that would ensure the stream quality decided upon, and prepare plans for legal enforcement of the standards.

It would have been logical to set standards for entire river basins since rivers don't always stay within state boundaries. What's more, there were already several regional river-basin compacts in existence that could have taken on the job. But with the single exception of the Delaware River Basin Commission, of which the federal government is a member, the government bypassed the regional bodies and insisted that each state set its own standards. Predictably, the result has been confusion. The states submitted standards by June 30, 1967, but by the end of 1969 Interior had given full approval to only twenty-five states and territories. It has now become the prickly task of the Secretary of the Interior to reconcile the conflicting sets of standards that states have established for portions of the same rivers.

Some states facing each other across a river have set different standards for water characteristics, as if dividing the river flow in the middle with a bureaucratic fence. Kentucky and Indiana, across the Ohio from each other, submitted two different temperature standards for that river: Kentucky came up with a maximum of 93° Fahrenheit, while Indiana wants 90°. Similarly, Ohio sets its limit at 93°, while West Virginia, across the same river, chose 86°. Up the river, Pennsylvania, too, decided on 86°. One reason for such differences about river temperature is that biologists don't always agree among themselves about safe temperatures for aquatic life. At one recent meeting in Cincinnati, where federal and state officials were attempting to reconcile the different figures for the Ohio, the disagreement among biologists was so great that one exasperated engineer suggested, "Maybe we should start putting ice cubes at different points in the river."

The biggest deficiency in the federal approach is its lack of imagination. Congress chose the subsidy route as being the easiest, but the task could have been undertaken much more thoughtfully. A regional or river-valley approach would have required more careful working out than a program of state-by-state standards and subsidies, but it would have made more sense economically, and would have assured continuing management of water quality.

A promising river-valley program is evolving along the Great Miami River in Ohio. The Great Miami runs through a heavily industrialized valley. There are, for instance, eighteen paper mills in the valley. Dayton, the principal city on the river, houses four divisions of General Motors and is the home of National Cash Register. To finance a three-year exploratory program for river management, the Miami Conservancy District, a regional flood-control agency, has imposed temporary charges, based on volume of effluent, on sixty plants, businesses, and municipalities along the river. These charges amount to a total of $350,000 a year, ranging from $500 that might be paid by a motel to $23,000 being paid by a single power-generating station.

With this money, plus a $500,000 grant from the Federal Water Pollution Control Administration, the district has been looking into river-wide measures that will be needed to control pollution even *after* every municipality along the river has a secondary treatment plant. (Dayton already has one.) The district's staff of sanitary engineers, ecologists, and systems analysts has come up with suggested measures to augment the low flood of the river as an additional method of pollution control. The Great Miami's mean annual flow at Dayton is 2,500 cubic feet a second, but every ten years or so it falls to a mere 170 cubic feet a second. To assure a more even flow, the Miami District will build either reservoirs or facilities to pump groundwater, at a cost of several million dollars. The cost will be shared by river users. District engineers are also exploring in-stream aeration, or artificial injection of air into the river, to provide additional dissolved oxygen. The state has set an ambitious goal for the Great Miami—to make the river usable "for all purposes, at all places, all the time."

To meet this goal, the district will introduce waste-discharge fees, which will probably be based on the amount of oxygen-demanding wastes or hot water discharged. Will these amount to a charge for polluting the river? "No," says Max L. Mitchell, the district's chief engineer. "Charges will be high enough to make industry reduce water use."

Federal money would do a lot more good if it were divided up along river-basin lines instead of municipality by municipality or state by state, with little regard for differences in pollution at different points in a basin. To distribute federal funds more effectively, Congress would have to overcome its parochial orientation. Also, Congress should be channeling more funds into new waste-treatment technologies and ways of putting them to use. Unless pollution abatement is undertaken in an imaginative and systematic manner, the "war" against dirty rivers may be a long, losing campaign.

II
Industry Starts
The Big Cleanup

by John Davenport

The disagreeable smell of hydrogen sulphide comes wafting through the magnificent stands of fir and cedar as a kind of early warning signal. Minutes later the highway leading up the northern shore of the Columbia River from Portland tips over a rise, and one looks down on a fortress-like industrial complex, which is wreathed in billowing clouds of steam and vapors worthy of Dante's *Inferno*. This is the great paper mill of the Crown Zellerbach Corp. at Camas, Washington. For years the mill's smoking stacks and chimneys have spelled jobs, opportunity, and prosperity for a land that a century and a half ago was wilderness—the chosen habitat of the Chinook Indians. Today that historic era of pioneering is taken more or less for granted. The big question is how and when Crown Zellerbach can produce paper without obscuring the view of distant Mount Hood, or disturbing the life of the Columbia River salmon, or affronting the nostrils of tourists, not to mention the citizens of this part of the Columbia basin.

This is a task that Crown Zellerbach is taking seriously, spurred on by multiplying state laws as well as by desire to protect its own corporate reputation as a good neighbor. "We have a big job to do, no question," says Francis

Boylon, president and chief executive officer. If you count it up, Crown Zellerbach is doing quite a bit. In 1968 the company completed a $3-million settling basin on Lady Island, off the Camas waterfront, which screens some 40 million gallons of water per day of their worst woody effluent. Now it is building a new $15-million magnefite pulping system, which will recirculate chemicals coming from its sulphite mill. Thereafter it will probably begin to revamp its kraft paper facilities at Camas to reduce their noxious sulphide odor—an objective already attained at its new mill at Wauna, eighty miles down the Columbia, and at its older Port Townsend plant on Puget Sound. In other areas of the country, notably at Bogalusa, Louisiana, Crown Zellerbach still has a considerable cleanup job to do. Yet counting in experimental work now in progress, the company is responding to a new kind of industrial challenge.

In doing so, Crown Zellerbach is representative of a much broader trend. In the Northwest there is growing sentiment that just as the lumber industry, after ghastly trial and error, learned how to use the forests without despoiling them, so the paper industry and others can meet the needs of an affluent society without drowning it in effluents. Throughout the nation, similar sentiments have been emanating from the leaders of the steel industry, the chemical industry, and public utilities. In 1969 industrial spending to clean up the air and the water, on conservative estimates, forged past the $1-billion mark. Meanwhile the manufacture of antipollution equipment —giant electrostatic precipitators for reducing dust and smoke, "scrubbers" for catching noxious gases, and plain old-fashioned valves and gates for controlling polluted water—has become a burgeoning industry in its own right. Companies like Research-Cottrell, Wheelabrator, Zurn Industries, and others have been engaging the attention of Wall Street analysts. On the same day that noted economists were predicting a "serious recession" for the economy as a whole, shares of the anti-pollution companies led the stock market upward.

All this is an immensely healthy development, because if the U.S. is to achieve a cleaner environment in the Seventies, industry has a critical role to play. Industry is only one contributor to pollution, but it is also the repository of the research and techniques that can bring pollution under control. "Technology—not regulations or good intentions or rhetoric—can preserve our clean water," says David D. Dominick, commissioner for the Federal Water Pollution Control Administration of the Department of the Interior. "Industrial leadership and political leadership must carry on the battle together."

Yet the battle is a curious and complicated one, and in defining industry's job it pays to maintain a sense of perspective and realism. "Ideas have their time in history," says one leading businessman, "and I don't think we should loose our screws about this thing." Analysis shows that the No. 1 polluter of the atmosphere is not the smoking factory or steaming chemical plant but the ubiquitous automobile, which at long last Detroit is trying to equip with antipollution devices whose costs will be borne by the consumer. When it comes to pollution of lakes and rivers, business is by no means the sole offender. Municipalities have been notably laggard in coping with their own raw sewage, and the effluents of industry that they have contracted to handle. If the shad find life uncomfortable in the Delaware River, it is not just because Sun Oil, Scott Paper, and Du Pont discharge wastes into that waterway, but because the citizens of Philadelphia and other cities have preferred all too often to complain about the pollution of the environment, rather than voting the necessary bond issues.

More important, it is unrealistic to think that business "leadership," however well intentioned, can by itself clean up the environment without clear guidelines from government, whose responsibility is to set the framework in which competitive enterprise operates. This conclusion is more than borne out by the survey of business executive opinion that appears in Chapter III. A clear majority of the chief executives who were canvassed believe that care for the physical environment is now a matter of urgent priority; 68 percent of the industrial companies polled have in fact set in motion antipollution measures—a remarkably high figure. Yet in most cases businessmen concede that they have mounted these programs in response to tightening regulation. A majority believe that Washington must go much further in unifying its regulatory policies and in providing incentives, whether through tax credit or other means, to spur investment in antipollution devices.

This is not surprising in view of management's fundamental responsibility to its shareholders. When a businessman invests in new equipment he at least knows that the resulting product will be paid for by the consumer, and he hopes it will yield a profit. But when he commits money to bettering the environment, he shoulders, on the near term at least, a dead cost; unless other companies follow suit he will find himself at a competitive disadvantage. This explains why many businessmen, who normally might be opposed to government regulations, have welcomed the Water Quality Act of 1965 and the Air Quality Act of 1967 as at least a beginning effort at setting national standards. Says Charles B. McCoy, president of Du Pont: "Everybody would be delighted if Du Pont or some other company could report that it has found a way to turn the expense of pollution control into an asset." With the present state of technology, he concludes, that just isn't possible.

Once these elementals are understood, industry's performance is seen to be neither as bad as sometimes made out nor as good as it should be. With respect to cleansing waste water, industry's expenditures have risen from about $45 million in 1952 to an annual current rate of about $600 million. Expenditures for protecting the atmosphere are somewhat smaller. But these figures do not reflect the full burden that many corporations are assuming. In many cases antipollution devices simply cannot be clamped on to existing facilities; instead whole plants must be redesigned with new productive equipment replacing old. Using a small sample, the National Industrial Conference Board estimates that investment in air and water pollution control rose from 2 percent of manufacturing capital outlays in 1967 to close to 4 percent in 1968. But a good number of companies queried by FORTUNE report that pollution control is now absorbing 10 percent of their capital budgets, and in two extreme cases the figure was 30 percent.

The great and significant fact is that, whatever the calculation, the figures are on the rise. Thus the private electric-power industry spent over $200 million in 1969 to counter air pollution, as against $127 million three years earlier. Its problems, both at the technological level and in terms of public relations, are representative of many other businesses. A high-energy economy requires increasing amounts of electrical juice to feed its factories and homes. Yet combustion of coal and low-grade fuel oil throws off tremendous quantities of particulate matter. In 1966 a survey of the Department of Health, Education, and Welfare estimated that power companies set some 2,200,000 tons of such matter adrift in the atmosphere. Equally serious, combustion of coal and to a lesser degree oil is a principal contributor to release of sulphur dioxide, which, while not toxic under ordinary conditions, has been implicated in most of the country's air-pollution crises (see Chapter VII).

Forced to operate close to the great metropolitan centers, and an obvious target for every housewife who glances out her window, the electric-power companies have moved to reduce pollution. Increasingly they have installed electrostatic precipitators, cyclonic collectors, and wet scrubbers to reduce emission of particulates, and in new plants such installations are now considered routine. Old and middle-aged plants cannot be abandoned, given recurrent power shortages and the danger of overloading and blackouts, which enrage the public even more than emission of soot and smoke.

Regulation is the only answer, and it is having its effect. Cincinnati Gas & Electric figures that its cumulative investment in antipollution equipment runs to nearly $8 million, and that expenditures will rise from 3 percent of total capital outlay to 5 percent and higher in coming years. Meanwhile operating costs for control of pollution shot up from $90,000 per year a decade ago to over $1 million in 1969, largely as the result of the company's being forced to use gas rather than coal in one of its older plants. In its area, use of low-sulphur coal is simply not practical.

Increasingly the power companies find that air-pollution control authorities in Washington and at the municipal level are increasing their costs, in a way that will force the raising of rates. State power commissions are committed to holding rates down. Caught in the middle, utilities have shown considerable ingenuity in shaving operating expenses. Southern California Edison and Arizona Public Service, plus smaller companies, formed a consortium to share the cost of pollution-control equipment at a huge generating station in New Mexico. Advances in technology are also in evidence. Kansas Power & Light Co. has shown that emissions of sulphur dioxide can be licked even in an old plant by a process devised by Combustion Engineering. Working closely with Pennsylvania Electric and later with Metropolitan Edison, the Monsanto Co. has developed a catalytic oxidation system that turns SO_2 into useful sulphuric acid. The cat-ox equipment is expensive, but sale of sulphuric acid can help in amortizing costs. As Monsanto salesmen are fond of pointing out to utilities, Monsanto can help to market this all-purpose byproduct.

A breakthrough in curbing sulphur dioxide would not only help the utilities, but also might rescue the steel industry from mounting public criticism. Oxides of sulphur are emitted in large quantities from steel's coking ovens, and are an old irritant in many a steel town. At its old Clairton works outside of Pittsburgh, U.S. Steel has turned to a cryogenic process that recovers some sixty tons of sulphur daily from material that used to go up the flue. Despite this reduction, the atmosphere around Clairton is far from savory, and an Allegheny County commissioner complained that Clairton "is the closest to hell I've ever been."

The steel industry also emits yellowish dust from its open hearths, and pollutes water with acid wastes from pickling steel as well as other operations. On one estimate the industry uses eight billion gallons of water per day for cooling and other purposes. Despite progress, it still has a considerable distance to go to obtain a clean bill of health. Pittsburgh's famous Golden Triangle is a kind of island rising from the polluted Allegheny and Monongahela rivers. Still worse is the condition of the Cuyahoga River, which snakes down into Lake Erie at Cleveland. Because of oil slicks and other contaminants, it is known as the "only body of water in North America that is considered a fire hazard."

Yet great and substantial changes are going on, and the lumbering steel giant is bestirring itself:

▶ Republic Steel is completing an $18-million waste-water treatment on a five-acre site in the Cuyahoga valley.

▶ U.S. Steel pioneered the use of electrostatic precipitators on open-hearth furnaces in the early Fifties. Its cumulative investment in air and water pollution control now totals well over $235 million.

▶ Bethlehem Steel is cleaning up the water it pours into Lake Erie at its Lackawanna mill, and its new plant at Burns Harbor on the Indiana dunes is a model of cleanliness. Over the next five years, Bethlehem figures that expenditures for pollution control will rise from 6 percent to 11 percent of its capital investment.

▶ Armco Steel poured $74 million from 1966 to 1969 into

air and water treatment facilities for its plants in Ohio, Kentucky, Missouri, and Texas, among other states, and it recently increased its maintenance budget by some $8,500,000. Says Harry Holiday, Armco's executive vice president: "All this has been and remains a considerable strain on our resources. The sums needed are often so big that no matter how much a company may want to be a good citizen, it's usually not so much a question of won't, but can't."

All told, the American Iron and Steel Institute figures that reporting members are currently budgeting over $325 million per year for pollution control of one kind or another. Technological change is on the march, notably in the matter of replacement of the open-hearth process by basic oxygen furnaces. As these are installed, precipitators, scrubbers, and water-treating facilities go with them. As of today, about 50 percent of all steel is produced by this process, and the conversion is bound to continue to meet competition at home, as well as from Europe and from Japan.

In this instance technological innovation, which conservationists often blame for pollution, actually helps to lessen it. The trouble in steel and metalworking generally is that old plants cannot be torn down overnight. Small and marginal companies lack the resources and the cash flow to keep up with multiplying regulation. A case in point is the financial condition of gray-iron foundries, which are usually small independent shops with comparatively small resources. "If you apply the same regulations to them as to steel," ruminates one expert, "three-quarters of them would simply go out of business tomorrow."

That would also have been true in the early days of oil. Initially refineries were spouting kettles of pollution, and had modern standards been applied to John D. Rockefeller's first famous still in Cleveland it too would have ignominiously folded to the discomfort of today's householder who might still be periodically rushing down to the basement to stoke his furnace with smoky coal. The thrust of technology and the rise of the great integrated oil company have changed all that. The oil industry may be held indirectly accountable for the automobile, which is the largest polluter of the atmosphere, but thanks to enormous expenditures—running to over $300 million in 1969—it has made an impressive start in cleaning up its refining operations.

In the San Francisco Bay area, for instance, Standard Oil (New Jersey) has loaded its new refinery at Benicia with the latest controls and for good measure has given it an aesthetic exterior. More remarkable has been the success of Standard of California in renovating its Richmond refinery, which nestles close to the city. The original installations here were made as far back as 1902 with little thought of the expansion needs of the future or the problems of contamination. Yet step by step, and often by small little-noticed improvements, the company has more than met required standards by use of precipitators; by extracting hydrogen sulphide from its refinery gases *before* they are reused for combustion (and selling it to a nearby chemical company); by waging unrelenting war on oil leakages; and by bulldozing out large oxidation basins where bacteria work on organic wastes and render them harmless before the water is swirled back to the sea. As if to reassure that stern conservationist, the National Audubon Society, lagoons beneath the cracking towers are the favored resting place of all manner of wildlife—gulls, avocets, godwits, curlews, and ducks.

Yet live ducks on San Francisco Bay will not quite make up for the dead gulls that strewed the beaches of Santa Barbara last year, as the result of the miscarriage of an offshore drilling operation by Union Oil Co. The great California oil spill has tarnished the industry's reputation and is all the more exasperating to many oil executives because in their opinion it was foreseeable and preventable. They point out that thousands of wells have been sunk off the California and Gulf coasts without mishap. Union Oil, it is felt, stretched the rules of the game by not sinking a well casing deep enough in a notably unstable geologic area; and the fact that it had permission to do so from the federal government doesn't mend matters. Clearly, individual companies will have to tighten up on their own drilling practices if they want to continue to exploit vital reserves. They will also have to cope with tanker spillage and with the threat of disasters such as the one that overtook the *Torrey Canyon* off the British Isles—a threat that grows with every increase in the size of tankers. Having made a start on cleaning up its refineries, oil faces a new challenge in production and transportation.

The chemical industry is more fortunate in that it generally buys its raw materials and feedstocks from others and so is free of the elemental production dangers pressing in on oil. Yet the explosive growth of organic chemistry and the use of petrochemicals to produce plastics, synthetic yarns, and a host of other products have given the chemical industry a new and less savory dimension. "Let's face it, we are still a dirty business," says Samuel Lenher, member of Du Pont's executive committee and now in over-all charge of its approach to the environment. Lenher is referring not just to the traditional throw-offs of chemical plants making inorganic commodities like sulphuric acid and chlorine. He is also referring to the complications that arise in making products containing the versatile carbon atom —intermediates for paints and dyes, and for the fibers that go into your suit or your wife's chemise. In these processes hundreds of different vapors must be scrubbed before being vented, and in the end myriad organic liquid wastes result. If not treated they will pollute river and lake just as much as a town's raw sewage.

To date Du Pont has made a cumulative investment of $125 million in air and water pollution control. Its yearly operating costs for this purpose run to over $25 million. Both figures are bound to rise as new plants are built and old ones renovated. Planning a new $10-million installation at Fayetteville on the Cape Fear River, in North Carolina, which will initially make Butacite for laminated safety glass, Du Pont went to extreme pains to bring in private and governmental experts before ground was even broken, and to provide facilities for secondary treatment of water through induced bacterial action. Installation of these and other controls will run to at least 10 percent of the plant's cost and account for as much as 20 percent of the engineering involved. The business is expensive but when undertaken at the start presents few technological difficulties, and production efficiencies may recoup part of the outlay.

What plagues Du Pont—and its experience is typical—are its old plants and specifically its great Chambers Works at Deepwater, New Jersey. Built during World War I, when the U.S. was cut off from the German dye industry, the works sprawl along the New Jersey side of the Delaware opposite Wilmington, in full view of 16 million motorists who annually traverse the Delaware Bridge. "We sometimes wish that bridge had been built downstream," says one Du Ponter, "because even steam these days is looked at with suspicion." The chief trouble with Chambers has been its industrial plumbing. When the plant was built water used for cooling was not separated from that used in the industrial process. Repairs are on the way, but meanwhile a big volume of waste flows into a single primary settling basin and from there is spewed out to the middle of the river.

That arrangement is far from pleasing to the Delaware River Basin Commission, which was set up in 1961 by Delaware, New Jersey, New York, Pennsylvania, and the federal government with broad powers over this polluted waterway. Headed by James F. Wright, a tough, outspoken ex-Navy officer, the commission has decreed that Du Pont must reduce the organic discharge of its oxygen-consuming waste load from 175,000 pounds per day to 32,000 pounds; it has imposed similar sweeping orders on other upriver plants including those of Sun Oil, Mobil Oil, and Scott Paper, as well as on various parts of metropolitan Philadelphia.

This would be a tall order if each party tried to go it alone. Instead, after considerable hassling, the companies and municipalities concerned have made a start on a cooperative scheme. Under it the commission has set up a pilot plant on land leased from Du Pont. If this proves successful, a huge secondary-treatment plant costing some $50 million will be built, though final financial arrangements still have to be decided upon. One proposal is that the commission should finance the plant through a public bond issue and then operate it, charging each company and municipality a user fee, based on the waste handled. In this case the companies would be spared the job of raising the initial capital, though their annual expenses will of course go up, under this or any alternative plans.

These developments going on in the dirty Delaware River basin represent a new and ingenious way by which companies can meet their obligations by a kind of pooling arrangement, and by which business and government can work in harness. Such joint effort, in the view of many experts, is going to become increasingly necessary as antipollution standards are tightened up. In the matter of air pollution, most companies will have to put in their own equipment at the plant site, paying for it as they go along. When it comes to water, a mixed pattern is emerging.

Big utility, steel, and oil companies such as we have been considering will for the most part continue to treat their own effluents. But food-processing companies, and thousands of small miscellaneous manufacturing firms, are already making large use of municipal water-treating plants. In 1968 about 15 percent of the waste water thrown off by all manufacturing establishments was sluiced into the public sewers of cities and smaller towns, and some 40 percent of the wastes handled by municipalities was industrial in origin.

Far from decrying this private use of public facilities, many engineers believe that it makes sound economic sense if full user charges are exacted. In some areas, notably California, municipal authorities have insisted that smaller factories use their water-treating plants on the grounds that this gives government real control over pollution. What is troubling is that the municipalities, even when duly compensated for handling industry's wastes, have been lagging behind in meeting their growing responsibilities.

Here, calculations made by the Interior Department are revealing. On one set of estimates the Federal Water Pollution Control Administration figures that between 1970 and 1974 industry must invest an additional $3 billion to take care of water effluents treated in its own plants. Since industrial investment for this purpose is now running to about $600 million per year, industry seems to be just about on target. By contrast, municipalities will have to invest some $10 billion in the same period to take care of rising population and industrial needs. But at present such investment is running to only about $1 billion per year or half of the total amount required annually.

This lag illustrates the fact that while almost everybody wants cleaner water as well as a cleaner atmosphere, there is much less acceptance of the fact that costs must be borne in one way or another. Industry can, and no doubt will, pass on its added costs through higher prices. It can also be helped by special treatment when it comes to amortization of equipment. But in the public sector the need for higher spending and higher taxes is inescapable. Yet the federal government has been most remiss. Washington has been very free with over-all antipollution legislation. It has been skinflint when it comes to voting the hard appropriations for the legislation's implementation.

In these circumstances a new force is fortunately coming into play, which will be exerting increasing pressure on government to do its job, as well as giving both government and industry better tools to work with. The makers of antipollution equipment are multiplying and have an increasing financial stake in seeing the whole drive for a cleaner environment succeed.

Yesterday equipment manufacturers were a small and

disorganized band of companies making air-screening devices for cement plants, and elementary valves and floodgates for factories or sewage-disposal plants. Today they are an increasingly vocal, chipper, and knowledgeable industrial group, which can be counted on to lobby for a larger and consistent effort, thus complementing the often angry and shrill outcries of the Sierra Club and other conservationists. The difference between the two groups is this: The conservationist looks at a smoking plant and in effect says: "Clean it up or shut down, come what may." The equipment-engineering firm says: "Look, you *can* clean it up, and we can show you the way."

The list of companies now so engaged is a long and diverse one, though their individual sales are still nothing to write home about. One of the biggest, Zurn Industries, which is incidentally the consultant on the Delaware River plan, had sales in 1969 of $109 million, of which about two-thirds were connected with environmental projects. Research-Cottrell, which is 100 percent committed to such business and claims to have built the first practical precipitator, had sales of $58 million. But the cumulative sales of both water and air equipment makers have been going up by perhaps 15 to 20 percent per year—a surge not unremarked by Wall Street.

The further encouraging fact is that all kinds of big corporations are getting into the act as a sideline. Joy Manufacturing Co. has its Western Precipitation Division and W. R. Grace & Co. is in the water business through the Dearborn Chemicals Division. Koppers in Pittsburgh has moved naturally from building coking plants into controlling their wastes; Westinghouse and General Electric and Alcoa have now seen that control equipment and related processes may be costly to buy but profitable to make and to market.

This emergence of environmental engineers should bring increasing pressure on Washington to provide incentives and money for a job barely begun. It may also lead to technological breakthroughs that ideally would allow society to turn so-called wastes into usable products. Monsanto's cat-ox process for trapping sulphur dioxide may be one such breakthrough. Another may be occurring in water treatment, where for years industry and city have relied on bacterial action to cleanse their effluents. In the view of engineers of Calgon Corp., a subsidiary of Merck & Co., this is a Model T method; and indeed it does seem strange that a society which can land men on the moon still relies on "bugs" to do its dirty work. Calgon is now pushing a new process using granular carbon adsorption to short-cut treatment of both human and industrial wastes.

What is significant is the new note of hope and realism that engineers are bringing to their task. Says Basil Welder, a director of marketing for Calgon: "It just isn't realistic to expect to catch trout off Battery Park . . . but I am convinced that the technology to handle water pollution is at hand." What's needed, in his view, is stricter enforcement of regulations now emerging on the books, plus full realization of the hard truth that, whether through taxes or higher prices, "costs must be passed on to consumers, who always pay in the end."

III

What Business Thinks About Its Environment

by Robert S. Diamond

No other group of Americans finds itself in quite as much of a quandary over the environment question as the men who head the nation's largest corporations. The decisions they make, and the financial resources they commit, will be crucial to success of the cleanup effort. They are under great public pressure to act responsibly, but on the other hand, they also have obligations to stockholders, employees, indeed to an economic system that thrives on ever increasing production and profits.

These sometimes conflicting pressures show up clearly in the results of a survey conducted for FORTUNE by Daniel Yankelovich, Inc. Some 270 chief executives of companies listed in FORTUNE's annual 500 directory were personally interviewed at length to ascertain their opinions about various aspects of the environment problem, as it affects them both as citizens and as leaders of business.

These chief executives are well aware that the environment has moved to the forefront of national concern. Six out of ten said the environment is a problem of "the highest priority," and is getting worse on virtually all fronts;

traffic, air pollution, congestion, water pollution, and garbage disposal were most frequently mentioned. When the executives were asked to rank the environmental question among ten issues, the average response of the panel as a whole was to put it about fifth in importance, well ahead of tax reform and health care, but well behind Vietnam, inflation, and law and order. As one Los Angeles industrialist put it: "We won't have to worry about pollution if we don't solve some of these other problems first."

Almost seven out of ten executives said they did not believe the environment is directly affecting their own health or that of their families, but a minority had strong feelings. A Pittsburgh executive complained about the putrid smell of sulphur from a nearby plant that wafts over his house each morning. A steelman is irritated by the noxious exhaust fumes that engulf his car as he sits idly in congested traffic every evening on the way home from work. An Ohio executive is distressed by the black ash that settles on his picturesque white frame house, keeping it in a perpetual state of untidiness. Reflecting the view of many others, a San Francisco executive said, grimly, "I'm aware of the condition of the environment daily and hourly."

When it comes to remedial action, a dominant sentiment is caution. The executives are concerned that government, in response to public pressure, will dictate the immediate spending of huge sums, which, rather than solve the basic causes of environmental problems, will sap the financial vigor of their companies. As a Cleveland executive put it, "We can't correct this thing overnight. We need a reasonable program, reasonably financed. We'd all go broke if we tried to do it in one year."

There is also concern that the public may not fully comprehend the extent of its responsibilities—and sacrifices—once the national commitment is ongoing and irreversible. "I don't think the American public at any level has really calculated the cost," says an executive in the rubber industry. "We have got into the habit of putting productivity increases into increased wages and fringe benefits.

The public has to be told that if we are going to clean up the air, the water, etc., this is going to cost money, and probably going to mean that from here out we are going to divert increased productivity to improving the quality of life rather than the quantity of possessions."

Like other citizens, the chief executives express some strong personal opinions on specific issues. More than half think that industry should spend more money to find an alternative either to the internal-combustion engine or to the automobile itself. Almost eight in ten favor some kind of effort to curb further population growth. But the building of the controversial supersonic jet is something else. Here the executives' pride in America's technological superiority overwhelms any other consideration. More than seven out of ten favor building the SST, though many who approve also indicate their anxiety over the possibility of window-shattering sonic booms.

The business leaders aren't nearly so certain about solving the environmental problems created by their own corporations, their plants and factories. They know they must do something, but how and how much puzzles and disturbs them. It may come as quite a surprise that the elite of business leadership strongly desire the federal government to step in, set the standards, regulate all activities pertaining to the environment, and help finance the job with tax incentives. Business, which has led the U.S. into unrivaled prosperity, wants in this case to be led. "I never thought I'd get to the point where I'd want the government to come in," said one executive, "but I don't think there's any other way."

The executives were asked two questions about the role of the federal government in the area of pollution and the environment. First, *would you like to see it step up its regulatory activities, maintain them at the present levels, or cut them back*? The responses:

Step up regulatory activities 57%
Maintain 29
Cut back 8
Not sure 6

Only in the South and among transportation and utility executives was there a notable reluctance to see the government step up its regulatory activities. Among the Southerners, most—53 percent—thought the government should maintain its activities at the current level, though 36 percent did favor a step-up. The same sentiment was generally reflected by the transportation-utility sector.

Second, *do you favor a single national agency to establish standards on air and water pollution control, land use, etc., or should local standards prevail?* The responses:

Single national standard 53%
Local standards 35
Not sure 12

Again, southern executives as well as those connected with transportation and the utilities took the position that local standards should prevail.

The way a large number of business leaders see it, substantial voluntary action on behalf of a single company is wasteful of corporate assets and ineffective in cleaning up the environment. They believe action on pollution must be collective. Otherwise, they argue, the costly efforts of a few companies won't even be noticed if corporate neighbors are still fouling the air and water. "We won't get this situation cleaned up except by laws that are enforced," said a Pittsburgh executive. "If I correct my plant problems, but my competitor doesn't, that company has a competitive advantage. I have committed huge sums; they haven't. In fairness to my stockholders, therefore, I can't make that first move." The executive added emphatically: "I see no hope except for legislation."

While conceding the necessity for government action, the corporate leaders face the prospect with some trepidation. They fear that standards will be established without sufficient study, only to be revised over and over again. As a Detroit official put it: "We don't want to be shooting at a moving target." The nervousness is predicated to some extent on the handling of the recent cyclamate controversy.

Some executives feel the government was excessively zealous and applied an extreme and unnecessary standard in arriving at the decision to ban all cyclamates. The government took that action when cancer appeared in test animals after they were given cyclamates in massive doses far beyond normal human consumption. In much the same manner, businessmen worry that the government will impose harsh standards, costly to implement, and possibly beyond what is necessary, to clean up America's towns and cities. Warns a mining executive: "The costs of reducing the last traces of air pollution—which may not be harmful—may be disproportionately more than it is really worth to society. In the last analysis, of course, the public will ultimately pay for it."

Many executives are worried lest pressure groups stir up a public alarm that could lead to extreme—and in their opinion—unwise measures being demanded of their companies. More than a few are suspicious of the academicians and conservationists who, they feel, do not appreciate the practical problems. The panel was asked: *Do you feel that the conservationist groups represent general public opinion or are they simply a pressure group?* The responses:

```
Pressure group ........................... 47%
Represent general public opinion ............. 38
Not sure ................................. 15
```

In some instances, the question drew angry replies from executives who feel that some—but not all—conservationist groups are using the currently voguish environmental issue to achieve political power. The San Francisco–based Sierra Club particularly came under attack. "I think that group has just gone crazy," said an Ohio executive. "If you listen to them we'd have no industry at all—just forests and streams." Describing a situation in which conservationists are attempting to block industrial development of an area on the Great Lakes, this executive declared: "They'd rather have sand dunes than huge new mills to serve the needs of the entire Middle West. That doesn't make sense to me." Another accuses the Sierra Club of raising money by making sensational accusations against big business. "They make news through unfounded charges they don't have to prove," said a New York executive. "They'll spend $100,000 for propaganda —like full-page ads in the *New York Times*—but not one cent to get the facts."

A series of questions bore on the judgments the executives might make when expenditures on environmental controls affect profits. *Should the protection of the environment be taken into consideration even if it means:*

	Should	Should not	Not sure
Inhibiting the introduction of new products	88%	8%	4%
Forgoing an increase in production	84	9	7
Reducing profits	85	9	6

Almost seven out of ten said their companies are participating in some type of antipollution effort. The question was: *Does your company participate in any community or industry-wide antipollution (or environmental-preservation) programs?* The responses:

Community	22%
Industry-wide	10
Both	37
Neither	26
Not sure	5

The executives were also asked whether their companies have a special antipollution budget or program of their own. Though, over-all, less than half (44 percent) said they did, it must be noted that bankers, insurance executives, etc., have no compelling reasons for establishing such a program. Among the nation's largest industrialists and retailers, 66 percent have company programs and spend money. But that figure drops to 51 percent among industrial and retail companies with under $1 billion in annual sales, and to 46 percent among utilities and transportation companies.

Of those companies that have antipollution budgets, an overwhelming number (81 percent) said they are spending more on a percentage basis than five years ago, and 63 percent plan to continue spending even more in their next fiscal year. It is still a small part of over-all expenditures; just over half said they committed 3 percent or less of their total capital budget last year for pollution control. FORTUNE asked: *What percentage of your 1969 capital budget was spent on pollution control?* The responses:

Less than 1 percent	19%
1–3 percent	32
4–5 percent	14
6–10 percent	15
11–15 percent	3
16–20 percent	3
More than 20 percent	3
Not sure	11

According to 86 percent of the executives, federal, state, and local regulations forced some of the spending. But a striking number of companies are spending funds for more than just stopgap purposes. Almost seven out of ten are allocating funds not only for present problems but also for research and development to prevent future pollution; among the largest industrialists virtually all of them are earmarking funds for research and development.

Forty-one percent of the panel concede that, in the past, corporate America accepted technological advances without adequately considering the consequences to the environment. "Until fifteen years ago, we accepted progress blindly," said one executive. "Now we are more and more conscious that technology must be accompanied by environmental control." Even so, some business leaders warn that it isn't entirely possible to see the ill effects of new technology in the making, regardless of the effort. "Things develop that you didn't expect," said one New York manufacturer. "DDT was considered a boon to mankind just a few years ago." Another philosophizes: "Tomorrow is dictated by what you learn today; our knowledge is evolving scientifically."

Even the most optimistic of businessmen believe it will probably take another decade—others, thirty years and more—significantly to turn the problem around. At the moment, however, business leaders aren't very impressed with the efforts of their peers. The panel was asked to rate the performances of eight basic industries, using a scale from 10 to 1 (poor). Here are the results:

Electric utility	6.0
Oil exploration	5.1
Steel	4.8
Detergents	4.7
Automobile manufacturing	4.6
Oil refining	4.5
Coal	4.5
Pulp and paper	4.4

As the results indicate, no industry gets very high marks. Only the electric-utility industry scored a clearly positive response. In all other cases there was general recognition that individual industries have done less than a satisfactory job.

To combat pollution, most (75 percent) agree that the major breakthroughs must be made in manufacturing and processing as opposed to basic product changes. But their research and development efforts cost money, and more than any other factor businessmen say that costs limit their efforts. Forty-one percent of the panel mentioned this, while 32 percent cited lack of technology. Presently the majority of businesses (58 percent) find the costs of fighting pollution cutting into company earnings, while others (24 percent) are passing on the costs to consumers through higher prices. Either way, most businessmen say financial assistance from the government in the form of tax credits would provide the best incentive. The question: *What do you think would be the single most effective—and least effective—incentives to business to do something more about pollution?* The responses:

	Most effective	Least effective
Tax credits for pollution-control costs	59%	2%
Industry-wide action	11	7
Government grants matching company expenditures	10	5
Government subsidies	5	15
Passing on costs to consumers	4	47
Improvement in the working environment	1	16

Curiously, many businessmen are somewhat sensitive about accepting government money. Rather typical is the case of a New York executive who found tax credits quite acceptable, but rejected the idea of government subsidies, saying: "No one wants to be in the relief line."

While conceding their responsibility, corporate leaders feel they should not be singled out as the only culprits. They point to the municipalities, which some of them believe do far more to pollute the waters than their own companies. They also think the public ought to do its part, too. Executives frown on the suggestion that their companies bear secondary responsibilities for such things as the prevention of litter caused, for instance, by tin cans and glass containers. Less than one executive in five believes his company has responsibility for the secondary effects of his business. "The trouble is people," declared a mining executive. "They can misuse anything. Hell, they won't even read directions, let alone follow them. If the directions say use a capful of detergent, they figure that if a capful is good a cupful is better."

Finally the panel was asked which American city has

the best—and the worst—living environment. The executives were easily in agreement on the worst—New York. There wasn't nearly as strong a consensus about which city has the best environment, although 69 percent of the panel mentioned western cities most frequently. The results:

Worst environment		Best environment	
New York City	61%	San Francisco	26%
Los Angeles	14	Denver	14
Chicago	6	Minneapolis-St. Paul	5
Newark	3	Phoenix	5
Detroit	1	Atlanta	5
Philadelphia	1	Dallas	4
Jersey City	1	Los Angeles	4

IV

The Economics of Environmental Quality

by Sanford Rose

Arguments over pollution control often exhibit the fallacy of all or nothing. Some people seem to feel that the environment should be restored to pre-industrial purity. Others apparently agree with the mayor of a smallish midwestern city who recently told a citizens' group: "If you want the town to grow, it's got to stink." Neither viewpoint is acceptable.

Those who imagine that pollution can be totally eliminated fail to grasp the dimensions of the waste problem. As some economists have recently suggested, it might be well to dispose of the expression "final consumption." People and businesses do not consume goods; they extract utilities from goods before discarding them. Such things as gems, works of art, heirlooms, and monuments might be thought of as being, in some sense, finally consumed. But all other goods—durables, nondurables, and byproducts—are eventually either discharged to the environment or cycled back into the production process. About 10 to 15 percent of total output, however, is temporarily accumulated in the form of personal possessions, capital goods, additions to inventory, etc. If society were to stop accumulating for a while, observes economist Allen V. Kneese of Resources for the Future, the weight of residuals discharged into the natural environment would equal the weight of raw materials used plus the weight of the oxygen absorbed during production. In other words, waste disposal would be an even larger operation than production of basic materials.

Though waste and pollution problems are too big to be eliminated, they can be ameliorated by producing fewer goods (or a different mix of goods), by recycling more of what has been produced, or by changing the form of wastes or the manner of their disposal. These alternatives are subject to economic evaluation. In principle, pollution is at an optimal level when the cost of additional amelioration would exceed the benefits. If by spending a dollar an upstream mill can save downstream water users at least a dollar, it should do so—from society's point of view.

Unfortunately, until very recently upstream mills were profoundly disinclined to spend anything to relieve downstream distress. They were accustomed to regarding the waste-disposal capacity of the stream as a free good. But wastes discharged upstream can impose costs of one kind or another downstream. Such costs are labeled, among other things, spillovers, side effects, external diseconomies, disamenities, and externalities.

One important effect of externalities is to warp the allocation of productive resources. Because a paper mill, for example, can get by without cleaning up its wastes, its costs of production are artificially understated. Since in a competitive economy prices tend to reflect production costs, the mill's prices may also be understated. If so, the result is greater demand for paper than if prices reflected the *full costs* of paper production—both the costs borne internally by the mill and those borne externally downstream. At the same time, some downstream producers— fisheries, perhaps—may have higher costs and prices, tending to depress demand. Society thus gets relatively too much paper and too little fish, and consumers of fish in effect subsidize consumers of paper. In some degree, then, resources are allocated with less than maximum efficiency and equity.

Externalities can be reduced (never eliminated) by many different means, including environmental standards, taxes, charges, subsidies, and generalized pressure. Each strategy results in a different mix of industrial, municipal, and federal expenditures on environmental quality.

Like any other investments, outlays for environmental improvement can be evaluated by standard benefit-cost analysis. Where capital outlays are required, as would usually be the case, the basic budgeting procedure is to forecast, for each year of the project's life, the probable benefits (damages avoided) minus the operating costs. Since it is a truism that a dollar earned next year is worth less than a dollar in hand, these net benefit levels must be discounted to their present value. Normally this is done by multiplying each year's net benefits by a discount factor based upon the estimated "opportunity cost" of capital—that is, what could have been earned if the funds had been used differently.

If the sum of the discounted net benefits exceeds the present cost of facilities, the project is economically sensible. If discounted benefits fall short of present costs, the project should be rejected (according to rational economics) because a dollar spent on further control would yield less than a dollar's worth of damage abatement—i.e., the marginal costs would exceed the marginal benefits.

In assessing the benefits and costs of pollution-control projects, government inevitably finds itself in a statistical scissors. Those that will have to pay for environmental improvements, such as businesses and municipalities, tend to inflate costs and deflate benefits. Those who particularly want the improvements—recreationists, let us say—can be counted on to do the reverse.

Many economists are convinced that of the two biases, the recreationists' happens to be the right one. Respectable project analyses, they argue, are typically biased in

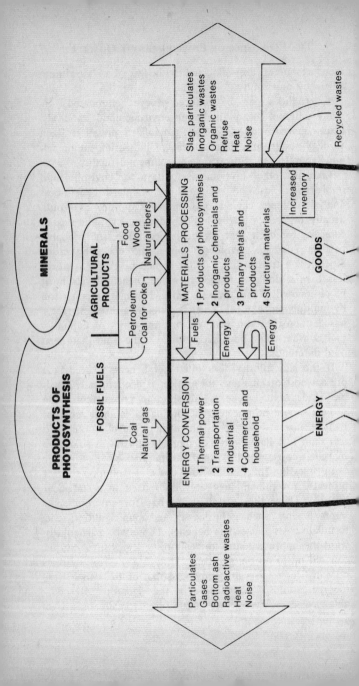

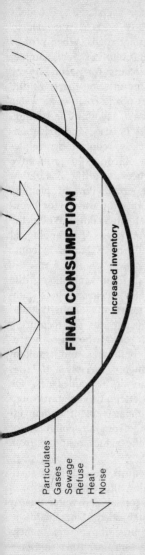

Particulates
Gases
Sewage
Refuse
Heat
Noise

FINAL CONSUMPTION

Increased inventory

You can't get rid of matter, according to a well-known law of physics. All you can do is transform it. Modern economies, like that of the U.S., are good at taking the concentrated and transforming it into the diffuse; they are not so good at doing the opposite. It is easy to turn coal into pollutants such as fly ash, gases, and soot, but difficult—economically, if not technologically—to turn the fly ash back into, say, cinderblocks. But we have to find ways to slim down those thick pollution arrows and fatten up that skinny recycled-wastes arrow. This diagram of material flows in the economy is adapted from a concept worked out by economist Allen V. Kneese and physicist Robert U. Ayres. Intermediate goods that are neither discarded nor used go into material-processing inventory, distinct from final-consumption inventory. The "final consumption" category embraces all goods that do not require further processing or assembly, regardless of who does the consuming.

favor of rejection, because costs are overstated and benefit
levels understated. Overstatement of costs is traceable to
the human tendency to travel familiar roads. When project
analysts talk about abating water pollution, for example,
they usually mean constructing plants for secondary or
tertiary treatment of effluents. But in some parts of the
country it would be much cheaper not to treat waste water
at all, but simply pipe it to storage lagoons for settling.
Eventually the waste water could be used for irrigation.
In other areas, costs can be greatly reduced by supple-
menting waste treatment with modification of productive
inputs, changes in production processes, artificial aeration
of streams, augmentation of low stream flow by planned
releases from reservoirs, and storage of wastes for even-
tual discharge during periods of high flow. When the
project analyst fails to scan the full range of technological
options—and he usually does fail—he is bound to come
up with an outsize price tag.

The benefits of environmental improvement, on the
other hand, tend to be understated because whole categories of damages are generally tossed out of the calculation. The researcher usually concentrates on measuring
physical damages. (This is hard enough because the re-
lationships between quantity of pollutants and resultant
damage are both complex and highly variable.) If he is
at all sensitive to the skepticism of his colleagues, the
researcher will make little or no attempt to quantify non-
physical damages—such as the impairment of human ef-
fectiveness or well-being resulting from air pollution, or
the loss of recreation and aesthetic values resulting from
water pollution.

Some respected pollution economists, indeed, have abandoned systematic damage estimation altogether. They point
out that governmental standards for air and water quality
are in the process of being established. Since damages
cannot be measured with either precision or comprehen-
siveness, these standards are in a sense arbitrary and sub-

optimal—that is, they may overcontrol or undercontrol pollution. But for better or worse, the argument runs, the standards are there, and economists should devote their efforts to working out least-cost ways of satisfying the new criteria.

As a result, most estimates of pollution damages are elaborate guesses. For example, the most frequently quoted figure for total annual air-pollution damage in the U.S.— $11 billion—is really an estimate of cleaning costs derived from smoke-damage data for Pittsburgh in 1913. In that year, investigators for the Mellon Institute calculated that Pittsburgh's smoke nuisance imposed additional costs— in cleaning and laundering clothes and maintaining and lighting homes, businesses, and public buildings—amounting to $20 a year per person. Many years later, this $20 figure was adjusted to 1959 prices on the basis of the commodity price index. The updated per capita damage estimate was then multiplied by the 1958 U.S. population to arrive at $11 billion.

One major omission from this obviously shaky figure is the cost of air-pollution effects on human health. In an as yet unpublished paper, economists Lester B. Lave and Eugene Seskin of Carnegie-Mellon University have made a serious statistical effort to assess this cost. Working from a variety of medical sources, Lave and Seskin correlated differences in mortality and disease rates in several geographic areas with differences in social class, population density, and two indices of air pollution. It turned out that air pollution and the bronchitis death rate were significantly correlated. According to these findings, reducing the amount of pollution in all geographic areas to the level of the cleanest region studied would lower the bronchitis death rate by between 40 and 70 percent. Other analyses revealed significant association between air pollution and heart disease, emphysema, lung cancer, and infant mortality.

Many of the medical studies that Lave and Seskin used

have been criticized on the grounds that the level of pollution may be correlated with unmeasured factors that are the real causes of ill health. For example, some of the studies do not control for differences in occupational exposure to pollution, in smoking habits, and in the general pace of life. Lave and Seskin took pains to try to meet this kind of objection. "It is especially hard," says their research paper, "to believe that air pollution and ill health are spuriously related when significant effects are found comparing individuals within strictly defined occupational groups, such as postmen or bus drivers (where incomes and working conditions are comparable and unmeasured habits ought to be similar)."

Lave and Seskin argue that roughly 25 percent of all respiratory disease is associated with air pollution. Therefore pollution must account for one-fourth of all the costs, both direct (mainly hospital and doctor bills) and indirect (basically, forgone earnings). Estimating conservatively, they say, the two economists find that the health damages attributable to air pollution amounted to $2 billion in 1963, the last year for which usable cost data are available. This is one of the biggest estimates to date. Ronald Ridker, a pollution economist now working for AID, put the 1958 costs of all respiratory disease at $2 billion (again, on a conservative reckoning) and the portion attributable to air pollution at about $400 million.

To carry out an adequate economic analysis of air-pollution abatement—or any undertaking to improve environmental quality—the analyst has to relate benefits to costs. More precisely, he has to estimate the specific dollar value of benefits to be derived from a given additional expenditure. A study done by Professor Thomas D. Crocker of the University of Wisconsin and Professor Robert J. Anderson Jr. of Purdue makes it possible to perform a marginal analysis of this kind in the air-pollution field. They argue that some of the damages associated with air pollution are reflected in property values. The potential home buyer perceives and evaluates many of the effects of air pollution on residential property—e.g., ailing shrubbery, off-color paint, sooty surfaces, unpleasant odors, hazy view, etc. And whether or not he connects such disamenities with air pollution, the buyer will take them into account in his offer price.

Anderson and Crocker tested their hypothesis statistically through an analysis of residential property values in three cities—St. Louis, Washington, and Kansas City. They correlated variations in sale prices and rents with family income, number of rooms, age and condition of property, distance from the center of the city, racial composition of neighborhood, and general educational level of neighborhood, as well as with two pollution variables—sulphur trioxide and suspended particulates. Air pollution and property values proved to be inversely related to a significant degree. Both pollution levels and property values differed in the three cities, of course, but in all three a moderate decrease in air quality—5 percent to 15 percent—correlated with a significant decrease in property values—from $300 to $700 for an average property. Roughly each 1 percent increase in either sulphation or particulates was associated with a .08 percent decline in the price or the rental value.

With these results, it is possible to calculate a damage figure for any city in the U.S. if certain information is available: median property value, the number of owner-occupied residential properties, and indices of sulphur trioxide and particulate pollution. To determine *annual* damages, the analyst multiplies the lost value by an interest rate to account for what the lost value would have earned if pollution hadn't wiped it out. In residential real estate a 12 percent rate of discount seems appropriate. Such calculations were carried out for eighty-five U.S. cities for 1965 (more recent air-sampling data were not available). The combined property-value losses worked out to $621 million.

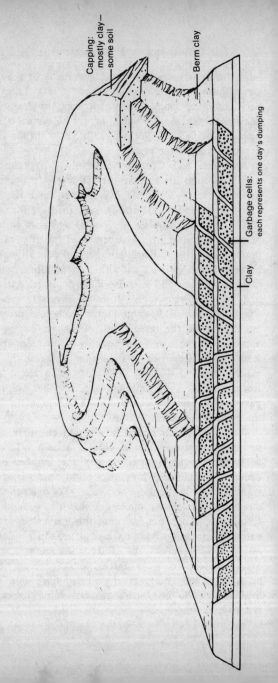

They'll ski down a mountain of garbage. Mount Trashmore, as it is frequently called in Du Page County, Illinois, is a hill slowly being sculpted out of garbage. Located in the Blackwell Forest Preserve, just west of Chicago, the hill will eventually reach 125 feet, the highest elevation in the county. It will feature six toboggan runs and five ski slopes in an area short on facilities for winter recreation. Mount Trashmore was conceived by John Sheaffer, a waste-management specialist at the University of Chicago. Du Page County officials came to him with two problems: they didn't know what to do with a badly scarred marshy pit, known as the Badlands, and they were running out of available landfill space for garbage. Sheaffer proposed excavation below the water table to turn the pit into a lake and use of the excavated clay to help build a mountain of garbage. The clay would form an impermeable barrier upon which garbage could be stacked without danger of groundwater contamination. The project, which started in 1965, is cheap:.the sale of gravel excavated from the pit will cover much of the cost. The mountain will be a honeycomb of cells, each four feet deep—three feet of garbage, one foot of clay. Each layer of cells is surrounded by a thick clay wall, or berm. County officials call their mountain, to be finished late this year, "the seventh engineering wonder of the world."

To complete the benefit-cost analysis, we need figures for the costs of canceling these damages. The National Air Pollution Control Administration has worked out some figures for the costs of controlling sulphur oxides and particulates in those eighty-five cities. The levels of control correspond to improvements in air quality of from 5 percent to 15 percent, according to Anderson, who helped prepare the NAPCA study. For fiscal 1972 (the first year for which NAPCA calculated a full set of cost figures), the low estimate of total costs came to $609 million.

So we have figures for 1965 damages and fiscal 1972 costs. To match them up, we must either deflate costs to 1965 prices or inflate 1965 property values to their projected 1972 levels. With either calculation, the marginal benefits of air-pollution control would far exceed the marginal costs. This conclusion, moreover, obviously understates the economic case for air-pollution abatement. Losses in residential property values encompass only a part, and perhaps not even a major part, of total air-pollution damages. Anything like a complete benefit-cost analysis—which would be an enormous undertaking—would clearly result in a decisive economic case for air-pollution control.

In a sense the water researcher faces a more formidable problem of damage estimation than Crocker and Anderson did. Property-value losses can be readily translated into dollar magnitudes, but a major benefit of water-pollution control would be more frequent and more satisfying outdoor recreational experiences—a commodity whose dollar value is estimated only with great difficulty. The researcher, in effect, has to assign what economists call a "shadow price" to a day's use of water recreation facilities. One way of doing this is to find out what the recreationist is paying for comparable services at private facilities. Another way is to use a method based on the cost of traveling to public recreational sites.

If the researcher can multiply a reasonable shadow price by the increase in recreational activity that would result from a given improvement in water quality, he can chart a curve of marginal benefits. If he can also determine what it would cost to achieve that improvement in water quality, he has a marginal cost curve. The point of intersection of the two curves represents the optimal improvement in water quality from the viewpoint of benefit-cost analysis.

This kind of study has been done for the Delaware River by economists Paul Davidson, F. Gerard Adams, and Joseph Seneca. What they considered on the benefit side was increased recreational use of the river for boating and fishing, specifically over the period 1965-90. On the other side were the costs of specified improvements in water quality—as measured by levels of dissolved oxygen —to make the river more suitable for boating and fishing. The three economists found that it took only very moderate shadow prices for recreation to justify the costs of the improvements. For example, it would have paid to clean up the Delaware considerably in 1965 if the use of the river for one day's boating was worth as much as $2.55 to the boater.

A shadow price does not imply that the users of the facilities would be required to pay it, though that is conceivable. The point is that if such a price is reasonable, enough additional social welfare (shadow price times extra recreational use) will be generated over the twenty-five-year span to repay the investment in river quality.

Some economists find fault with the Delaware River study on the ground that too low an interest rate—5 percent—was used in discounting future benefits. While this rate is higher than the 3 to 4 percent typically used for public projects, it is much below realistic rates for the private economy, even in 1965. The use of 5 percent means, in effect, that funds worth, say, 10 percent in the private sector would be diverted and put to work in public projects at about one-half the yield. In straight economic terms this is plainly inefficient. Says George S. Tolley, professor of economics at the University of Chicago: "I believe in letting the chips fall where they may. If a public project cannot stand on its own it should not be buttressed by artificially low discount rates. In fact, if proper discount rates had been used, many of the dams and reservoirs built over the past few decades would never have been started. They couldn't have paid their way."

If the three economists had used a more "realistic" discount rate, they would have had to raise the shadow price for boating in order to justify cleaning up the Delaware to the specified levels of water quality. Since $2.55 was probably a bargain for a day's boating even in 1965, they have some room to maneuver. What's more, increased recreational use is obviously not the only benefit from a cleaner Delaware River. When a river is upgraded, riparian property values rise dramatically. In proper benefit-cost accounting, each such subsidiary benefit should be taken into account.

Recreational benefits and higher property values are also important in benefit-cost analysis of solid wastes—garbage, trash, ashes, sewage sludge, building rubble, auto hulks, beer cans. Sometimes called the "third pollution," solid wastes are normally regarded ·as a nuisance, or worse. They can also be regarded as resources—resources out of place. The general objective of solid-waste control might be thought of as the displacement of wastes from locations where they have negative value to locations where they have positive value.

One means of doing that is the sanitary landfill. Untreated waste is buried daily in layers, each covered under several inches of compacted earth. The landfill technique has transformed thousands of acres of low-value land into parks, playgrounds, golf courses, bathing beaches, marinas, parking areas, and other useful facilities. While it is sometimes said that the future of the landfill is limited by land scarcity, there is actually plenty of low-value land to be improved. Karl Wolf, head of research of the American Public Works Association, points out that strip-mining operations mangle more than 150,000 acres each year. If this ruined land were used for landfill, it might be enough to hold our entire annual output of solid wastes.

In areas where the landfill is surrounded by residential or commercial property, transitional problems arise. Although completed landfills improve property values, the actual filling operations temporarily impair values. People living near an active landfill have to contend with refuse-truck traffic, noise, odor, and unpleasant views. The longer these disamenities persist, the more basis property owners have for demanding compensation. Unfortunately, large landfills may take quite a few years to complete. And community landfills tend to be large, partly because they stir resentment; politicians, wanting to limit the number of angry voters, prefer one big site to several smaller sites.

One way to speed up completion of large landfills would be to pool wastes over extensive areas. This approach, of course, would require the cooperation of independent municipalities. It would also require some inexpensive means of long-distance haulage. The railroad seems the likeliest answer here. Since many different transport operations are required for large landfills—truck collection of waste, transfer to train, retransfer to trucks, and dumping at the landfill site—it may even make sense for the railroads to become vertically integrated waste disposers.

Landfill can be thought of as a form of recycling—in this case, using wastes as a construction material. Wastes may have higher value as fuel, and some wastes still higher value as industrial materials. Quite a bit of trash and garbage is incinerated in the U.S. today, but in ways that change wastes from one form to a possibly even less desirable form, air pollutants. It is possible to design efficient incineration systems that emit little pollution and yield useful heat. In Europe utilities burn trash to provide steam, but this use of solid wastes is relatively rare in the U.S. It might be more economically attractive if refuse vehicles were designed to pick up tanks of used oil from service stations and spray their contents on household garbage as it is collected, thereby raising the BTU content of refuse.

Because there is a pervasive throwaway psychology in the U.S., we do not come close to realizing—or even envisioning—the potentialities of recycling. In many instances where recycling is dismissed as economically or technically unfeasible, the possibility has not been carefully examined. The steel industry nowadays recycles much less scrap than it used to because the basic oxygen furnace, unlike the older open hearth, supposedly cannot assimilate very much cold scrap; industry leaders argue that heating costs make recycling uneconomic. Research scientists at M.I.T. disagree. They point out that the basic oxygen furnace produces a lot of hot waste gases. These could be passed up a tower filled with pulverized steel scrap, which could thus be heated enough for recycling.

An increasingly troublesome category of solid wastes in the U.S. is the detritus of what has been called "the packaging explosion." In packaging, we have moved away from recycling instead of toward it. The returnable beverage bottle of yesteryear, a casualty of affluence, has given way to the throwaway bottle and the throwaway can. The can, what's more, is not rustable steel anymore but persistent aluminum. In an affluent democracy, it is not feasible either to make people stop littering the landscape with beer cans and similar artifacts or to pay other people to go around picking them up. But it should be possible to arrange matters so that a lot more used containers get recycled. Two problems are involved here—to induce people to return containers to the retailer and to make it worth the retailer's while to bother with the things.

The prospect of getting back a small deposit is for most people these days a weak inducement. The popularity of lotteries suggests a more effective alternative: some kind of arrangement by which the return of every five-hundredth or thousandth soft-drink bottle or beer can pays the lucky returner a windfall of, say, $5 or $10. Where would the money come from? Partly from the recycle value of the containers and partly from taxes applied by society at appropriate points.

One kind of tax that might be imposed is a levy on containers, graduated according to the difficulty of recycling. There always seems to be some perverse component that gets in the way of economical recycling. The aluminum ring that the twist-off cap leaves around the neck of the bottle makes it uneconomic to grind such bottles into cullet for glassmaking; the cost of removing the metal, either before or after grinding, is too high. The small magnesium content of the aluminum can lowers its salvage value. And the tin coating and lead solder on the otherwise steel "tin can" largely exclude it from the salvage market. Unless this obstructive heterogeneity can be dealt with, society is unlikely to make much of a dent in what might be called "the beer-can problem." If containers do not have a worthwhile salvage value, the retailer will balk at assuming the costs involved in handling them—they would become *his* waste-disposal problem.

In some measure, government could help through discriminatory procurement policies. Federal purchases account for about 6 percent of total packaging expenditures. If government were to insist upon tin-free steel cans or magnesium-free aluminum cans, industry would be powerfully prodded to alter the technology. Taxation, however, would provide a more comprehensive approach. To be effective, such a tax would have to be high enough to provide those lottery-type payoffs *plus* subsidies to cover the extra costs of putting the salvaged material into recyclable forms. If the tax were well designed, it would encourage the manufacture of more recyclable containers. Ideally, the design of the tax should also encourage degradability—not all containers would be recycled in any case, and it is desirable that litter be as evanescent as possible. Some cheerful scientists even look forward to the "biodegradable beer can," which would hold together on the grocer's shelf but succumb to bacterial decay after it was thrown away.

The disposal tax seems to be on the right track in that it makes the polluter assume costs related to his pollution. In the water and air pollution fields, proposals similar to the disposal tax are usually called "effluent charges." The concept of the effluent charge is beginning to get a friendly hearing in Congress. Senator William Proxmire of Wisconsin introduced a bill to place an effluent charge on water polluters. Although no figures are presented in the bill, Resources for the Future, which helped Proxmire prepare the legislation, has suggested a charge of 8 to 10 cents per pound of biochemical oxygen demand (BOD) added to any waterway. If a levy of this kind were imposed nationally—and some think companies might relocate if it weren't—national revenues at the 8 to 10 cent level would come to around $2 billion to $3 billion. Part of the revenues could go to finance regional development authorities that would undertake research and plan and construct collective treatment facilities.

Advocates of the effluent-charge approach argue that it would operate far more economically than the present arrangements for water-pollution abatement. Currently the federal government attempts to set—or get states to set— waste-treatment standards requiring large industrial and municipal expenditures, and provides some financial assistance, generally to municipalities. These standards often call for excessive uniformity, in that every producer in a given river basin must abate by the same percentage. Such a program is bound to be inefficient. For one thing, it requires small plants with high marginal cleanup costs to treat to the same level as large plants, which usually have lower marginal costs. For another, it ignores locational factors and stream hydrology—i.e., the obvious circumstance that a polluter in one part of a stream can discharge wastes without causing any damage, while another polluter elsewhere can cause a lot of damage with the same amount of wastes.

The effluent charge remedies at least the first deficiency, because each water user along the stream can find his own least-cost mix of pollution-abatement measures and effluent-charge payments. Moreover, an adequately designed system of effluent charges gives the polluter an incentive to carry treatment further than he would under a uniform standard. If the standard is 85 percent removal of biochemical oxygen demand, the polluter will presumably not exceed it. But if he has to pay for all BOD, he has an incentive to push treatment to a higher level, say 95 percent removal, as long as the additional cost is less than the effluent charges avoided.

The superiority of the effluent charge over the uniform standard has been documented in a 1967 study of the Delaware estuary by Grant W. Schaumburg of Harvard. He found that if proportional abatement were required in the Delaware, the area's forty-four major polluters—industrial and municipal—would pay a total of $106 million to clean up the water to a specified level of dissolved oxygen; but under a reasonable system of effluent charges, the combined cost would be only $61 million. Further cost reduction could probably be achieved by placing entire river basins under the jurisdiction of regional authorities that would build and administer large-scale treatment facilities. These can yield substantial economies of scale. Economists Andrew Whinston and Glenn Graves reckon that if four polluters on the Delaware, each with a waterflow of 2,500,000 gallons a day, built individual treatment plants to remove 50 percent of BOD, the cost of all four together would come to $1,520,000 a year. But if the four were located close enough to each other, they could become partners in a single plant and save a total of $480,000 a year.

Whether Congress selects the effluent-charge route or

not, it seems clear that the public is of a mind to devote more of the economy's resources to environmental improvement. It is impossible to predict with confidence how much the coming campaign to reverse environmental deterioration will cost, or what the impact will be on the national economic accounts. With the aid of an econometric model, however, it is possible to explore this largely uncharted territory.

Econometric models of the U.S. economy can be used to estimate the effects of an altered investment mix resulting from the imposition of additional pollution controls. Choosing a model that had predicted changes in the U.S. economy with reasonable accuracy during 1962–64, economist Robert Anderson altered some investment and price inputs to reflect a fairly stringent level of air-pollution control. He then reran the model for the two-and-a-half-year span from the first quarter of 1962 through the second quarter of 1964. Anderson assumed that (1) manufacturing industries were required to increase their investment on air-pollution devices at an annual rate of $1.2 billion; (2) public utilities were required to increase *their* outlays at an annual rate of $320 million; and (3) new-car prices rose by 1 percent owing to the installation of emission-control devices.

What happened to G.N.P.? It went down. At the end of the two and a half years, that is, G.N.P. was at an annual rate of $617 billion, whereas without the assumed pollution controls the model had predicted more than $625 billion. Unemployment was up: the new rate worked out to 5.3 percent instead of 4.8. And prices were up too—the price level was about 1.2 percentage points higher than had been projected.

For the first four quarters of the simulation, G.N.P. reached higher levels with the air-pollution controls than it would have without them—but largely because of price effects. In the very short run, demand for most goods tends to be "price inelastic"—that is, most people don't change their buying habits quickly as prices rise. As a result, industry's increased costs of coping with air pollution could temporarily be passed on to consumers in the form of higher prices, which are, of course, reflected in G.N.P. But after a while the decline in disposable income tends to stiffen resistance to price increases. Demand weakens and output starts going down.

This exercise in retroactive simulation suggests that increasing expenditures on pollution control would produce some of the effects of a recession. But for this kind of assessment, econometric models are biased on the side of pessimism. For one thing, the model is incapable of taking technological change into account. In this case, that is an especially serious shortcoming. The spectrums of technological options for pollution control are likely to expand quite dramatically, even in the short run. Abatement technology is just in its infancy. Since society has for so long allowed free-of-charge use of the natural environment for waste disposal, incentives to do research in abatement have been weak at best. As a result, pollution-control equipment tends to be relatively inefficient, and the potential trade-offs among waste treatment, process changes, and recycling are poorly understood. Once society—by one means or another—begins charging rent for use of the environment's capacity to absorb wastes, engineers will have to think about pollution control as an integral part of plant design rather than as an afterthought. A lot more research funds will be allocated to pollution control, and costs may go down faster than anyone expects.

It is even conceivable that industrial costs will not rise at all in the medium or longer term. Pollution control not only provides incentives for more efficient operation through recycling, but also makes cities more livable. And people who work in more livable places don't have to be paid quite so much as those who work in less livable places. If wage rates in the future are just slightly lower than they would have been if the cities had remained polluted, the difference might quickly offset industry's increased pollution-control costs.

Over the longer term, pollution abatement seems likely to *increase* real G.N.P. A significant decrease in air pollution, for example, can be expected to reduce absenteeism and turnover and improve productivity. Some industries, perhaps many industries, might have to pay out less in sickness and death benefits. With turnover reduced, they might also have lower training costs. If these longer-range savings were put into a thorough benefit-cost analysis, many corporations might discover that pollution control yields a profit, entirely apart from any altruistic considerations.

In a broader sense it is a mistake to put any great emphasis on the G.N.P. aspect. Although the costs of environmental improvement are reflected in national product, many benefits are not. For example, when property values rise because of a decline in air pollution, the community's real wealth or capital stock increases; but this shows up in G.N.P. only to the extent that actual or imputed rents go up or real-estate salesmen's commissions get bigger. Similarly, although environmental improvements may enrich leisure and so increase satisfactions (or reduce dissatisfactions), these benefits could not be reflected in G.N.P. at all, as G.N.P. is presently reckoned.

Moreover, it is difficult to know what is really "net" in gross national product since the measurement counts both output and some of its adverse consequences—every good and service that has a price. To arrive at even an approximate gauge of the annual increase (if any) in welfare, we would have to deflate G.N.P. by a magnitude that might be called G.N.E.—gross national effluent. G.N.E. would be a statistical basket for all those negative goods and services produced in the course of, or as a result of, the production of positive goods and services. Negative goods and services in this sense include additional transportation to escape the effects of environmental impairments, additional cleaning, additional medical services, goods prematurely replaced because of soiling or corrosion, and, of course, pollution-control equipment. If we subtracted G.N.E. from G.N.P., the remainder would be a better measure than G.N.P. of what the economy has done for us in any year. And it is certain that pollution control will sharply increase the value of this remainder.

V

Cars and Cities on a Collision Course

by Allan T. Demaree

As a blessing, the automobile is far from unalloyed. In many U.S. cities, motor vehicles contribute as much as 75 percent of the noise and 80 percent of the air pollution. They voraciously consume great quantities of land for parking and highways, breaking up neighborhoods, uprooting residents, and displacing businesses. They clog downtown streets and freeways until, as one British observer put it, "buildings seem to rise from a plinth of cars." They browbeat pedestrians and jangle nerves. They abet that homely, unplanned urban growth called sprawl, which some fear will permanently turn America into an endless string of Tastee Freezes. They kill and injure more Americans annually than a hundred Vietnam wars. One transportation expert, Wilfred Owen of the Brookings Institution, wonders "whether it is possible to be urbanized and motorized and at the same time civilized."

Many Americans are familiar with—even support— these indictments. Yet individually they cling devotedly to their cars, cherishing the independence and convenience that comes from owning their own wheels. And therein lies a dilemma that threatens to become far more painful. For the great factories in Detroit are adding to the car supply at a rate of 22,000 a day, and the men who make cars confidently predict that production will climb to 41,000 a day by the end of the decade—the greatest ten-year increase in history. Even after allowing for the junking of old vehicles, the American Automobile Association expects the number of autos, trucks, and buses to climb to 170 million by the year 1985, over 60 percent more than are on the road today. To handle traffic requirements by then, state highway officials estimate more than 40,000 additional freeway miles will be needed.

For public consumption anyway, some of the automobile manufacturers still lightly dismiss the significance of current objections to their product: "I don't feel it's a crisis," says James M. Roche, chairman of General Motors. "We've talked about congestion for a long time. Back in 1929 people said that no more cars would be sold because there weren't enough roads to handle them."

But Henry Ford II takes a different view. He freely concedes that the rising production of material goods, including Ford cars, has "been purchased at a high cost in environmental pollution—dirty air, dirty water, ugly landscape." One Ford executive has warned that "social and political pressures for curtailing automobile use and promoting other forms of transportation will continue to mount." The Ford Motor Co., therefore, is reacting with the instincts of an entrepreneur, hoping to take advantage of the revolution in public expectations and design systems —such as automated highways—that will better serve the needs of the cities and, in the words of one official, "represent an orderly transition from Ford's current products."

The impact of the growing flood of cars will be felt most keenly in America's bustling cities. By the end of this decade, metropolitan traffic volumes are projected to increase roughly 40 percent in Pittsburgh, 50 percent in Boston, 90 percent in Detroit, and 100 percent in Los Angeles. In the manufacturers' home city of Detroit, planners predict that traffic will be moving slower and slower in the future, despite regional expenditures of $3 billion for new highways and arterial roads, and $1.1 billion for a rail rapid-transit system. "If you're going to drive," says a Detroit planner, "you'd better do it before 1975."

For cities, cars are not only unaesthetic and disruptive, but wasteful as well. Between 250 and 300 square feet of space is needed for every car that commuters park in the city. Cities typically devote 10 to 20 percent of their downtown land solely to parking cars these days. A single lane of city freeway carries a maximum of 3,000 people an hour in cars, given the average occupancy rate of 1.5 persons per automobile. By contrast, buses and rail rapid transit using the same amount of space can transport as many as 30,000 and 40,000 persons respectively. Most highway interchanges built to accommodate auto commuters take up at least forty acres of land and sometimes twice that much. Mayors rankle at seeing freeways push taxpaying properties off the tax rolls.

Resentment against the omnipotent car is growing. Methods of subduing the machine are being diligently sought, in the disorganized and seemingly uncoordinated ways Americans make progress. Many proposals are quite modest, such as better control of traffic flows or encouragement for commuters to use new mass-transit services. Others are visionary, like building automated highways to guide cars through congested areas in great volumes at high speeds. Still others seek better to plan the community's physical layout—where people live, work, and play —so that travel demands can be anticipated and systematically met. No single solution provides a panacea, and the combination of tactics adopted will vary widely in different cities, depending on such critical factors as their size and age, their topography, their rate of growth, their residential densities, and a host of other factors.

One obvious way for a city to defend itself is to fight off new highways. A series of freeway revolts began in 1959 when San Franciscans rebelled at the ugly intrusion of the double-decked Embarcadero Freeway and forced the road builders to discontinue it, leaving the highway literally suspended in midair. Since then San Francisco has turned down hundreds of millions in federal highway funds rather than allow the city to be violated. Revolts have followed in Baltimore, in Washington, D.C., Indianapolis, Cleveland, Philadelphia, New York, and elsewhere. In Boston last December, Mayor Kevin H. White cited the "anguished objections of neighborhood residents" and urged the governor of Massachusetts to order "an immediate halt to any land taking, demolition, or construction . . . for new highways."

But there is a price to pay for such blocking tactics. The traffic keeps coming on relentlessly, and in the absence of freeways to handle it, the jams worsen.

The root cause of all the trouble remains untouched. Few possessions have become so intricately entwined in the fabric of American life as the car. The same car that has proved awkward and ill-mannered in the city made possible the single most important movement in postwar living, the massive migration to the suburbs. And, for those who can afford it, the car serves suburban living well. It frees the driver from reliance on fixed schedules, from waiting for buses or trains in cold or rainy weather, from lugging packages in crowded, uncomfortable public conveyances, and from annoying transfers from one form of travel to another.

America's fierce allegiance to the automobile was strikingly revealed by two nationwide surveys in 1967 that posed this question: "The auto pollutes air, creates traffic, demolishes property, and kills people. Is the contribution the auto makes to our way of life worth this?" Four out of five respondents answered yes, even in metropolitan areas where the adverse impact of the auto is greatest.

The decline of public transit is yet another gauge of consumer preference for the car. More than 250 transit companies have discontinued operations in the past fifteen years, leaving some cities and towns without service. Where there is service, it tends to be so primitive, and equipment so old, dirty, and lacking in amenities, that prospective passengers are repelled. Thus, as cities seek a substitute for the automobile, they find themselves battling established trends of consumer acceptance.

If one failure stands out above all others in leading us to our present state of affairs, it is that we haven't paid adequate attention to what might be called the "demand side" of the transportation equation. The physical arrangement of where people live, work, and play has a momentous impact on their demands for travel. In a typical British new town, for example, the average trip to work is only a couple of miles. In Detroit it is nearly nine miles. One reason for the disparity is that nearly every American city has concentrated on increasing the supply of highways, rather than controlling the demand for them by strict zoning and land-use planning.

Until fairly recently, the zoning in Chicago would have allowed the entire U.S. population to be housed in that city, and at one time New York was zoned for some 370 million people. Houston still has no zoning law at all. "We build a freeway expecting the area to be residential," says a Texas highway official, "and when it's completed, shopping and office centers spring up. It's impossible to plan."

Under these circumstances, it is little wonder that cities fail to reserve the open space and parkland that is necessary to meet recreational and aesthetic needs. But this failure feeds back on auto travel with disconcerting effects. When weekend traffic barely crawls to Jones Beach on New York's Long Island or Fregene Beach outside Rome, the fault doesn't necessarily lie with the automobile and the roadway. It may be that we have too few beaches or inadequate substitutes for them, such as community swimming pools.

Major metropolitan areas are now doing some better planning. They all established planning commissions in the early Sixties, as a result of federal legislation requiring them to draw up comprehensive transportation and land-use plans in order to qualify for federal highway funds. The commissions' reports, which often run to three or more thick volumes, describe which areas should be developed for residential, industrial, and commercial use, and which reserved for park and recreation land. Then they prescribe a transportation network, whether highways, rapid transit, or both, that will meet the needs of the community efficiently.

The trouble is that the metropolitan regions are powerless to carry out their plans. Large metropolitan areas comprise scores, even hundreds, of independent local governments that have no obligation—and often no intention —of following the plans in matters of zoning and land use. If a developer proposes a shopping center that would enhance a village's tax base, the village government is likely to approve the project, regardless of whether adequate roads exist to serve it. Until something is done to put power in planning, metropolitan areas will continue to sprawl incoherently, creating a constantly changing pattern of transportation demands that can never be adequately met.

Far more attention has been focused on the "supply side" of the transportation equation. Almost universally, large cities feel the need to provide more mass transit, even in the face of transit's historic decline and the heavy financial burdens that transit imposes. Undoubtedly, this trend toward what is commonly called "a more balanced transportation system" is in the right direction. But the movement is beset with confusion and turmoil.

The goals cities hope to achieve are so multitudinous— and even conflicting—that they are frequently confused by city leaders and misunderstood by the electorate. The administrators seek to enhance the mobility of the poor, to revitalize downtown as a retailing center, to decongest access routes to the central business districts, airports, and other centers of activity, to reduce the amount of land consumed by highways and parking, as well as to cut down air pollution. Implicit in these goals is the need for difficult value judgments. Is it a good investment of limited community resources to sink money into rapid transit rather than schools or recreational facilities? If rapid transit is proposed as a means of revitalizing a flagging central business district, might it not be wiser to let downtown decline naturally and quicken the dispersal of its functions to the suburbs?

Choices necessarily discriminate among different classes of citizens. A system that best serves commuters from the suburbs may prove seriously deficient in meeting the needs of the central-city poor. Rail lines can provide good, economical commuter and downtown-distribution service where the demand for transportation through fixed corridors is very high. Five U.S. cities—New York, Philadelphia, Boston, Cleveland, and Chicago—have such transit systems, and two others, San Francisco and Washington, D.C., are building new ones. San Francisco's Bay Area Rapid Transit, which is expected to begin operations by 1972, will require an annual subsidy about double its fare-box revenues, and will carry mainly suburbanites to the commercial centers of San Francisco and Oakland. This will certainly reduce traffic and lessen the demand for more highways. But it will do little for the poor and autoless, who remain near the city's core. The line passes through few low-income areas, and where it does the stations are so far apart that few poor people will be able to use it without driving or taking a bus to a station.

Equally perplexing is the fact that transit systems must be hand-tailored to the communities they serve. Consider, for example, the problem of carrying commuters into the central cities of Boston, Milwaukee, and Houston. Practically the only thing the three cities have in common is that they all face the threat of growing congestion unless mass transit is provided to complement the automobile.

Boston, an old city whose narrow streets were laid down more than two centuries before the advent of the auto, built the first subway in the U.S. (in 1897) and has a long history of heavy transit patronage, both rail and bus, although this patronage has been declining in recent years. Peak-hour travel into the already congested core city is expected to increase about 10 percent by 1975. The downtown street system cannot be altered to accommodate this volume in cars. The number of people that would have to be dislocated, the heavy investment in existing buildings, the very history of the place foreclose that possibility.

So Boston's civic leaders hope that the number of peak-hour transit riders will increase 20 percent while auto travel edges up only 3 percent. The city is expanding and improving its rail system, which seems a wise choice under the circumstances. Boston's population density, among the highest in the nation, generates enough patronage to employ efficiently rail's high hauling capacity. Moreover, the fact that the city already has large investments in the subway, which can be regarded as sunk costs, makes improving that system more sensible than building a new rail line would be in a city of comparable density.

Houston is at the opposite pole. While Boston houses more than 14,000 persons per square mile, Houston's residential densities are less than 3,000, hardly enough to support a rail commuter system. While 45 percent of the travelers to and from Boston use public transit, in Houston only 4 percent do, and a glance at the queues at bus stops shows that 4 percent to be predominantly the black and the poor. The local bus company was so scantily used that it went into receivership in 1966 and has since been taken over by National City Lines. That company has put the line back in the black, but has been unable to increase ridership despite innumerable experiments. Two new routes, a dime-a-ride shopper's special and a twenty-mile run to the Houston Intercontinental Airport, both lose money and the latter will probably be discontinued.

So the task of getting Houstonians out of their cars and onto public transportation is a particularly vexing one, though the city leadership sees it must be solved if downtown isn't going to be swamped with cars. Traffic in and out of downtown is expected to triple by 1990 and, as city planner Roscoe Jones puts it, "When you start putting freeways between freeways, there's no city left to go to." A recent study recommended that Houston adopt some form of rapid transit to downtown. Some experts think nothing short of a futuristic "personal transit" system carrying passengers over a net of automated guideways would induce Houstonians to give up the car.

Milwaukee falls between the extremes of Houston and Boston. Early in the century it was well served by a network of electrified interurban rail lines so snappy that when commuters lit their cigars, they struck their matches on serrated German silver pieces mounted by the windows. As elsewhere, however, the auto's siren song was heard: in 1951 the railroad ceased passenger operations. Now the local bus company is losing ridership to the point where the county is talking seriously about a government takeover.

One bright sign runs counter to the general decline in transit. In 1964 Milwaukee's bus company began an experimental service that sped suburban riders nonstop over freeways into the city, cutting a fifty-six-minute ride down to thirty-three. This service has grown and prospered. Two out of every five riders formerly drove their cars into Milwaukee. Now they park in fringe lots provided free by suburban shopping centers where the "freeway flyer" routes begin. The Milwaukee region is planning to build an exclusive two-lane busway along the most densely traveled corridor to the city, probably by paving over an east-west right-of-way that was formerly used by the electrified rail line. The busway is expected to be congestion-free, carrying 47,000 riders each weekday by 1990, or 25 percent of the rush-hour travelers who commute along that busy corridor. This would mean a reduction of about 6,000 in the numbers of cars that might otherwise be driven into Milwaukee and parked.

In many urban areas like Milwaukee, buses appear to be headed for a renaissance. Since they can carry ten times more people per lane than cars, buses use existing highways more efficiently and lessen the need to send more roads slashing through densely populated areas.

Buses also have important advantages over fixed rail systems. They can operate at 60 mph over private rights-of-way, then return to the street system to pick up passengers or distribute them downtown near their jobs. This flexibility means buses can come closer than rail to rivaling the automobile's fast, convenient, door-to-door service. Moreover, buses require far lower capital investments than rail, which makes them less costly in all but the few major cities that can fully use rail's extremely high capacity. (The exceptions are cities of very high density, most of which already have rail systems or are building them, and cities like Seattle and San Francisco, where topographical features channel the population into dense corridors that fixed-rail lines serve well.)

Today, the very word "bus" connotes a crowded, fuming, dirty, uncomfortable monstrosity whose seats bear evidence that the younger generation still knows how to wield a penknife. Buses don't have to be that way, of course. General Motors has developed what it calls the RTX (Rapid Transit Experimental), which has such amenities as lounge seating, comparable in spacing to first-class on an airplane, a lower floor for easier entry, larger windows, and carpeting. The bus will use a gas turbine engine that gives smoother acceleration, more efficient power, and most important, a substantial reduction in pollutant emissions. Production of such a bus is not expected before 1972 at the earliest, however, and its cost will undoubtedly be higher than current models.

Although proponents never advertise the fact, building a special right-of-way for buses or providing them with an exclusive freeway lane is a subtle resort to coercion. It snatches lanes away from auto drivers and gives them to bus riders, pressuring people to leave their cars at home. Early experiments, however, have managed to keep from stirring up drivers by giving the buses newly built lanes not previously used by cars. The first experiment to devote special lanes to buses on an interstate highway began last September, when two new lanes along a four-mile stretch of the congested Shirley Highway were given to buses heading into Washington, D.C., during the morning rush. Free from congestion, the buses clipped twelve to eighteen minutes off the time cars took to make the same run.

Advocates of exclusive bus lanes say they are giving people preference over vehicles. But their plans can actually militate against the "people preference" principle if the bus lanes are under-used. In the Shirley Highway experiment, for example, there are five Washington-bound lanes, including the two new ones allocated to buses. The two bus lanes carry some 2,300 passengers during the morning rush while the three auto lanes carry 10,000. This means that 40 percent of the roadway is currently devoted to buses carrying only 19 percent of the commuters. The hope, of course, is that drivers will see the faster moving buses and switch, but the extent to which this will happen is not yet clear.

A more sophisticated method of improving highway capacity without risking under-use of special lanes is to meter all vehicles onto the roadway, and then give buses preference. In the metering system, vehicles are stopped by traffic lights at freeway ramps, and are prevented from entering if they would overtax the freeway's capacity and cause congestion. By keeping the highway free-flowing, the number of cars that move over it can be increased by as much as 15 percent and speeds can be raised 20 percent. Although several freeways are already flow-controlled, none of them has the bus-preference feature, which would significantly increase their capacity to carry people.

Another bus-oriented innovation is designed to increase the mobility of people without cars and, at the same time, supply a service so convenient that it will attract people who might otherwise drive. The system goes by many names, dial-a-bus, DART, CARS, taxi-bus, and others, but the concept is basically the same. It is designed to give

customers door-to-door service in maneuverable twelve-
to fifteen-passenger buses that can be summoned by telephone like a taxi. When a customer calls for service, a
controller at the bus company punches the request into a
computer. In a fraction of a second the computer selects
which bus to assign to the new customer, and then fits that
customer into the bus's schedule. The system is custom-
tailored to the kinds of travel demands inherent in urban
sprawl, where destinations are so widely dispersed that
they cannot be reached by conventional buses and rail. It
could provide an ideal feeder service, collecting passengers
in residential areas and taking them to rapid-transit buses
or trains bound for downtown.

Daniel Roos, an M.I.T. professor who has designed such
a system, believes it can serve suburban areas which generate between twenty and 100 requests for service per
square mile every hour, a level of demand that falls below
the breakeven point for conventional buses. The costs for
the computer and control equipment for a system of 100
buses would come to no more than 20 percent of total
costs; fares charged could be expected to run about one-
third that of a taxi and half again as much as a conventional bus. To find out whether enough people will respond
to the service to make it economical, M.I.T. is seeking
federal funds to put the first dial-a-bus system into operation.

Many astute transportation experts believe that a far
better solution is within grasp, which uniquely combines
the amenities of the automobile with rapid transit's high-
hauling capacity. One such system is called dual-mode.
It would employ cars similar to today's, but they could
operate in two modes, either manually driven over conventional highways or under automatic control over specially
designed guideways through heavily traveled corridors.

On the guideway, all driver functions would be per-
formed automatically, with the vehicle picking up its
guidance and speed instructions from a third rail. The
driver could read or relax until his exit. The car could

have two power plants, an electric motor to pick up power from the third rail and an internal combustion engine for propulsion on regular highways. The essential advantage of the system is that automation would allow vehicles to travel just a few feet apart at speeds up to 60 or 70 mph. A single guideway seven or eight feet wide could carry the same number of cars as five lanes of highway, 10,000 vehicles (or 15,000 people) per hour.

The guideways are relatively lightweight steel structures that could be prefabricated and erected over existing rights-of-way, such as highway medians or railroad tracks. Ford analyzed a possible dual-mode system for Detroit and found that all but eight miles of it could be constructed over rail rights-of-way. Users could drive from their homes to a guideway, enter, and travel by automatic mode to their destinations. There they could either leave the guideway and park, or if the area were congested—or if public policy dictated—they could leave their cars on the guideway to be taken automatically to an out-of-the-way storage area.

Ford is gathering a consortium of four major corporations bent on developing the dual-mode idea with private capital. Foster Weldon, director of Ford's transportation research and planning, sees dual-mode as an ideal "transition from Ford's current products," since the car is altered mainly by the addition of control equipment, including a retractable arm that would attach to the guideway. Milwaukee County—in concert with American Motors, Allis Chalmers, and Dwight Baumann Associates—is drafting a request for $10 million or more in federal funds to demonstrate a dual-mode system on a mile-and-a-half guideway. Officials see dual-mode as an evolutionary step beyond the busway they have planned, and if the experiment is successful, they would consider extending the guideway to residential areas and the city's downtown.

No modest extension of current technology, the dual-mode system is still fraught with unknowns. Commuters who see the hazards of ordinary short-haul train travel daily increasing are likely to be skeptical about the whole dreamy-sounding proposal. General Motors shied away from developing dual-mode because the potential damages in case of an accident would be so great that special legislation limiting liability would be needed. (G.M. is also under an antitrust consent decree barring it from entering into mass transit as long as it produces buses.) The fail-safe mechanisms necessary to assure safety at high speeds bumper-to-bumper have yet to be developed and proven. Moreover, building elevated guideways, even though they are lightweight, might bring ugliness and blight reminiscent of the elevated railway.

Even with the most vigorous expansion of public transit imaginable, the number of automobiles operating over conventional roads is expected to keep right on climbing. Transit systems may alleviate rush-hour congestion, but they will not substantially lower the over-all demand for auto travel. In St. Louis, rapid-transit proposals are expected to reduce peak-hour traffic on certain highways in 1980 by 25 percent; but total motor vehicle trips are expected to decline only about 5 percent.

So the need to better fit highways into the communities they serve is not slackening. The modern highway has been rightly impugned as a concrete monster, a big ditch, or a Chinese wall. A six-mile stretch of roadway in Houston is nineteen lanes across. The depressed and depressing Dan Ryan Expressway in Chicago is fourteen lanes wide, an uninspiring view for those who dwell in the equally dismal public housing along the way.

To avoid such divisive effects, a research team from M.I.T. has proposed that highway builders negotiate a new road's location and design with interested community groups. This is not an approach highway engineers are accustomed to, for they have generally considered it unprofessional to scratch around in parochial politics. Because engineers have tended to ignore the highway's impact on communities it penetrates, they have frequently been subjected to what Marvin Manheim of M.I.T. calls "the big surprise." They study the highway location, run benefit-cost analyses, propose a route publicly, and then are surprised by the overwhelming community opposition it creates. In the approach suggested by M.I.T., road builders would discuss with members of the local communities various possible routes for the highway, alternative plans for relocating those displaced, compensation for their losses, and ways to wrap into the highway design plans for enhancing the neighborhood with parks, schools, or recreation areas.

The California State Highway Department has already set a notable precedent for this kind of sensitive highway planning. In designing the Century Freeway through Watts, where 2,600 families were to be uprooted, the department seriously analyzed the suggestions of community groups, rather than confronting them with a *fait accompli* at a public hearing. When it came to a final choice between two freeway corridors, the California Highway Commission chose the one favored by the residents over the one proposed by the highway department staff, even though this meant substantially higher expenditures. The commission also stipulated that no resident would be left worse off as a result of the highway, either financially or in the quality of his housing.

This took imaginative planning. Most of the homes in the path of the freeway were single-family dwellings, half of them owner-occupied, averaging $13,000 in value. Under normal procedures the highway department would simply pay the fair market value for the property and the occupants would be forced to find housing elsewhere. Most of those displaced had their roots in Watts and wanted to remain, yet mortgage money there was practically non-existent as a result of the riots. Comparable replacement housing outside Watts would cost about 50 percent more than residents would have received for their old homes.

The Highway Department seized the opportunity to use the freeway investment, more than $100 million, not just to build a road, but to rejuvenate housing in Watts. It won authorization from the state legislature to acquire scattered vacant lots in the area roughly six blocks from the freeway and move single-family homes onto these sites. The Watts Labor Community Action Committee, a highly respected self-help group in the area, is training local labor in preparing the home sites, moving the houses, and re-habilitating them. Significantly, the program is tailored to the desires of the Watts community, where people want single-family homes, not high-rise apartment buildings, which the highway builders had originally proposed.

With staggering growth in traffic volumes expected from coast to coast, many cities will be finding it necessary to find innovative approaches, which accommodate travel without degrading the environment. The search may some day resolve the conflict of interest between car and city.

VI

How Baltimore Tamed the Highway Monster

by Judson Gooding

The automobile's challenge to the heart and the life of American cities has rarely been posed—or met—as squarely as it has been in Baltimore. The city was suffering from traffic paralysis, especially crippling because as a port city its economy relies heavily on the movement of goods. As a cure, old-line highway engineers devised a traditional solution funneling the entire flow of local and interstate traffic right through the city's center. But a group of Baltimore's leaders, concerned about the social and aesthetic effects of the proposal, managed to block it. They have sponsored a new approach, designed by architects—the first major effort of its kind in the country—that promises to enhance and enrich the city, rather than to gouge an ugly scar through it.

Baltimore had uneasily acknowledged a need for major improvements in its traffic accommodations as far back as the 1940's. At that time, New York's controversial promoter of public works, Robert Moses, was brought in as a consultant. Moses recommended an expressway traveling east and west through a long-suffering district now known as the Franklin-Mulberry corridor. The fact of his recommendation alone accelerated a spiraling deterioration in the neighborhood, and gave Baltimoreans their first realization of what such a road could do to a city. Franklin-Mulberry slowly turned into a slum, despite its neat row houses with their white marble stoops. A blockwide strip has now been razed to make room for the road, but even today, more than two decades after Moses' recommendation, no work has been done on it.

Over the next twenty-four years, indecisive officials allocated a staggering $15 million to studies of different routes and plans for mass transit. Meanwhile the immense, overpowering 42,500-mile, $60-billion interstate highway network had hurtled up to Baltimore's borders at top speed; I-95 barreled up from the south and down from the northeast; I-83 bored downward from the north; and I-70N drilled in toward the city's center through the expanding western suburbs. The engineers were pressing to go ahead. The cars were already there, honking and teeming. But many influential Baltimoreans saw the confluence of interstate highways as a massive, multipronged concrete shaft aimed straight at the city's heart.

The road that threatened Baltimore was labeled 10-D. Plans for it were developed by a consortium of engineers led by the J. E. Greiner Co., Maryland's dominant engineering firm and a political powerhouse whose directors have over the years established an exquisite rapport with state highway officials. Designed so that it would channel traffic through the city's center, 10-D would have bisected a stable, prosperous Negro neighborhood called Rosemont. It would have taken up 160 acres of vital downtown land for a vast and ugly interchange; it would have set a fourteen-lane bridge across the city's old inner harbor, endangering its redevelopment; and it would have bitten off portions of two historical neighborhoods, Fells Point and Federal Hill. The road would also have wiped out 4,000 dwellings, taken $28 million worth of land off the tax rolls, and ended 6,000 jobs. Its harshest effects would have fallen on the poor.

In short, it was a highwayman's road, designed to meet the complex federal interstate regulations, produce the

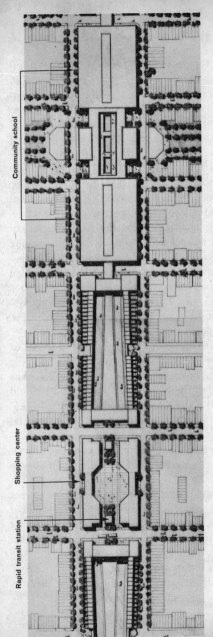

Rapid transit station

Shopping center

Community school

Refuting the belief that roads must destroy, Baltimore's expressway will go underground at some points, permitting the use of the space above for community facilities. This sector of the expressway, which runs submerged for a mile and a half through the Franklin-Mulberry district, will include a school, covering a three-block area above the road, and shopping centers. The design provides for a rapid transit line between the lanes of the expressway. Since the expressway is in part a kind of underpass, local streets can easily cross it. This should help maintain neighborhood cohesion.

most paved surface for the least cost, and take the shortest feasible route from point to point. It was an expression of the American subservience to the car, which concedes that the car can and should go wherever it wishes, with no questions asked about the effects or alternatives.

Opposition to 10-D was crystallized by a Baltimore architect and planner, Archibald Rogers, fifty-two. Rogers worked through the local chapter of the American Institute of Architects, which was already distressed by the aesthetics of the Jones Falls Expressway, Baltimore's first and only fling at expressway building. He won the chapter's support for his contention that the road should not mutilate, but should instead actively serve the city, making a positive contribution to its structure, its economy and—extraordinary thought!—to its beauty. Rogers suggested that the over-all design responsibility not be the exclusive province of the engineers, but be given to a "design-concept team" combining engineering, planning, sociological, and architectural expertise.

Enthusiastic support came from David Barton, head of the Baltimore planning commission, which controls the city's master plan and can make or break any public-works proposal. "I jumped on the bandwagon," he says. "I would have jumped even if it had been a garbage wagon—anything to get away from the 10-D plan." Later, crucial encouragement also came from an influential city councilman, William D. Schaefer, who blocked passage of condemnation ordinances needed to clear the right-of-way for 10-D.

The uprising against the expressway was to some extent like those that have occurred in San Francisco, where a freeway along the bay was stopped dead in its tracks, and in New Orleans, whose historic French Quarter was preserved from freeway intrusion by citizen action. But the Baltimore revolt was different in that it offered its citizens an alternative, a realistic option that would allow them to have the road they needed, but a road that would be responsive to their wishes and to many of their nontransportation requirements.

Archibald Rogers expressed the aspirations for better design in his July, 1966, proposal for a concept team. The objectives, he wrote, were "to design an efficient, safe and beautiful urban freeway system as a well balanced and organized entity related visually and functionally to the surrounding urban physical fabric, both existing and planned." This meant varying the freeway so it related to the different sorts of areas it was to traverse—depressing the road and putting a park over it, where a district needed recreational facilities, or lifting it above ground level where it went through an industrial area which needed warehousing space beneath the road, or building housing over the road in residential districts where housing was in short supply. The intention was also to provide for coordination with other means of travel, particularly with future mass-transit facilities, so that the systems would complement each other; parking lots and garages would be built at transit junctions so motorists could leave their cars and walk or take trains, reducing road congestion. The road would be routed around established, stable neighborhoods and districts with historical or scenic values, rather than slicing through them. Another important aim was to make the road as aesthetically pleasing as possible, treating it as a public monument of grace and beauty rather than a mere sluiceway for disposing of cars.

The State Roads Commission, which through its Policy Advisory Board was responsible for the planning of the expressway, asked Rogers who might best head the concept team. He suggested Skidmore, Owings & Merrill's senior partner, Nathaniel A. Owings, sixty-seven, who was then head of the presidential commission on Pennsylvania Avenue. "I called him up," Rogers says of Owings, "and he went for it like a salmon goes for a fly."

The hook Owings leapt for was alluring, but barbed. It offered him a remarkable chance to develop new highway concepts for a major city in ways that had never been tried before. But there were hazards built in. The interstate highway's points of entry into the city's fabric seemed impossible to change. Practically all of the condemnation ordinances for the land along the proposed 10-D route within the city boundaries had been passed, and were almost irrevocable, since once land has been doomed in this sort of proceeding, irreversible deterioration sets in. Owings also faced the open hostility of the engineers, who cherished their own proposed route, and insisted that he be prohibited from propagandizing either federal officials or the Baltimore public. In addition there was the demanding deadline imposed by the imminent expiration of federal funds for the interstate system.

Still, Owings set to work, in his words, "to lace tubes of traffic through vital parts without unduly disturbing the living organism of the city." In order to ensure that the expressway would improve the community, public and private "joint development" construction and action programs were to be undertaken in concert, all along the twenty-two miles of freeway corridor. It was intended that the $600 million to be spent for the expressway should serve as a catalyst for inducing private and public agencies concerned with housing, parks and recreation, as well as industry, to invest the same amount along the corridor.

When the $4,800,000 planning and engineering contract was signed, the architects opened a Washington office, with partner John Weese in charge. Joining in, at first as subcontractors, was the consulting-engineer firm of Parsons, Brinckerhoff, Quade & Douglas, and traffic consultant Wilbur Smith & Associates, both of which had carried out major transportation studies for Baltimore. Consultants in community relations, landscape design, sociology, acoustics, and other disciplines were also signed on.

The concept team started with a $60,000 study to ascertain the need for a controversial part of the 10-D system, which would lead the heavy volume of traffic on I-95 into the city from the southwest. When it reported that the leg was indeed necessary, the tensions eased somewhat between the concept team and the automobile fraternity—truckers, highway engineers, and builders, the whole vast automotive-sales-service-gasoline business complex—which resented sharing power that had traditionally been theirs alone. The highway engineers feared that they would be relegated to the role of mere draftsmen, and they echoed the engineering fraternity's criticism of "petunia-planting aesthetes, bird watchers, and do-gooders" who give little thought to cost and utility. They not incidentally disliked seeing highway design fees, once all theirs too, split several ways.

During the power struggle over the extent of the concept team's mandate, Jerome B. Wolff, then Maryland State Highway director, and Bernard L. Werner, at that time Baltimore's director of public works, publicly announced they were having "very distinct differences of opinion" with Owings and implied he might be fired. "He will accept our dicta . . . or we don't think he can properly be a part of the expressway program."

The engineering firms relaxed their opposition to the concept-team idea only after Transportation Secretary Alan Boyd had accepted it, at the urging of Federal Highway Administrator Lowell Bridwell, and agreed that the government would pay 90 percent of the planning costs. The fact that the engineers would share handsomely in the big $4,800,000 planning contract fee was a significant factor in their grudging acceptance. The contract finally worked out created a group called Urban Design Concept Team, a joint venture that included the original participants plus the J. E. Greiner Co. Since Greiner was thus given the status of partner, participation in the concept team became even more acceptable to the engineers. Despite this concession, the architects retained control over the urban design, and the engineers took over only after the basic plans were made.

The Skidmore, Owings & Merrill staff teams and consultants carried out depth surveys of neighborhoods in the path of the expressway, to learn what effects the road would have. As the findings came in, Owings moved to enlarge his franchise. His sociologists, social psychologists, and housing economists found that Rosemont, the Negro residential area with a high percentage of employment and owner-occupied houses, would be cleaved in two by the expressway. Owings sought authorization to study an alternative route.

That brought loud objections from Highway Director Wolff, who said that it would be "the first and the last change" to be made under the design team's limited mandate. Nonetheless, the Owings team eventually won the argument over Rosemont, and succeeded in reducing the bad effects the expressway would have had on Fells Point and Federal Hill. It also completely eliminated the downtown inner-harbor bridge and 160-acre interchange. The designers contrived a bypass route leading through traffic around the south end of the city through industrial areas, thus enhancing road service to an important district that until then was difficult of access.

Carrying out its studies, the concept team held talks with residents throughout the city. The people in the Franklin-Mulberry corridor expressed desires for housing, industry, schools, play fields, community facilities, and shopping centers. But it soon became distressingly clear that they did not want the new interstate route at all. "Expressways are for people with cars," one black man said. Mrs. Esther Redd, executive secretary of the Relocation Action Movement, founded to help find housing for displaced residents, said: "The whites will use the road to get to their high-paying jobs in the city, while they live out in the country and pay low taxes. The poor will be pushed out of their houses."

Such reactions led Joseph Axelrod, chief of the interstate division for Baltimore of the State Roads Commission, for the moment at least to prohibit any further community work by the concept team. He says the multidisciplinary team approach "is a brilliant concept for the designing of a highway, or for any public work" and he praised the "multi-design approach, incorporating the function of the work, with economic, social, and city needs woven in." However, he believes that "too many of the concept-team people are too close to their ivory tower," and criticizes some who became "emotionally involved." The ultimate resolution of the difficulty remains uncertain.

Despite this setback, the concept team went on to attack the proposed 10-D route, believed to be locked in both by the terms of the contract and by the condemnation ordinances. The team's study showed that the interchange and inner-harbor bridge would be constantly overburdened from the day they were opened for business, the overload increasing as traffic volumes grew. The survey also showed that 43 percent of the traffic to be funneled through the city's center was not headed for downtown, but was through traffic.

Consequently, the team developed two alternative routings, with a bypass, suggested by landscape and highway consultant Michael Rapuano, through industrial areas south of the downtown center. The bypass would allow narrowing the expressway routes through the city, leaving more land in the condemned strip for joint development, and lowering the volume of traffic. The two alternative designs differed chiefly in that one more closely resembled 10-D, keeping a smaller inner-harbor bridge crossing (3-C), while the other eliminated it entirely (3-A). Both provided boulevards for central-business-district circulation and traffic access into downtown areas via spurs off the expressway. But Greiner Co.'s Edward Donnelly still preferred the old favorite, 10-D.

Finally, Baltimore's Mayor Thomas D'Alesandro III, who had authority to make the decision, called a climactic meeting of concept-team members and all the city and state officials concerned. There, he listened to lengthy arguments for 3-C, which were really tantamount to support for 10-D, with Owings dissenting in favor of 3-A. Finally he said: "If all of you here are in favor of 3-C" —and he paused, while Owings' heart sank—"it's got to be wrong. I am adopting 3-A, and I don't want to hear any more about the difficulties, I want to hear how it will be done." Longer by 4.4 miles because of the bypass, 3-A will cost more than 10-D: $600 million versus $384 million. But the social and transportation gains seem worth the cost.

Now the long years of controversy appear to be ended. The local papers and the community have accepted the decision with relief, if not with total unanimity. The engineering profession and the civil-service highway functionaries still nurse a forlorn hope that 10-D can be brought back to life. But most Baltimore observers are sure 10-D is dead.

Whatever happens, the concept-team approach has been tested in a major project under heavy fire, and has proved valid. The team's contract has been renewed, a crucial turning point, with another $4,700,000 recently approved by the Department of Transportation. Detailed planning of the fourteen segments into which the 22.4 miles of roadway have been divided is going ahead. Concept-team members are working with private developers and with the Greater Baltimore Committee, a civic development group, to advance a number of joint development projects. They could total $600 million or more if fully implemented.

In Baltimore at least, the automobile is no longer undisputed sovereign over the land.

VII

Some
Burning Questions
About Combustion

by Tom Alexander

Civilization has been formed in the crucible of fire. Combustion, or the rapid chemical reaction of oxygen with carbon, has propelled human advancement. But Prometheus' presumptuous gift of fire to man is calling down a punishment not anticipated by the ancients. For where there's fire, there's smoke.

Fossil fuels that accumulated over hundreds of millions of years are being converted to gas and ash in a combustive gluttony that began a century ago. In the U.S., the tonnage of these fuels consumed doubles every twenty years. All this combustion—the internal combustion of transportation and the external combustion of industries, power plants, home heating, and incineration—is by far the principal contributor to the dirtiness of cities and the foulness of air. Still, if pollution is the brother of affluence, concern about pollution is affluence's child. Air pollution is not a recent invention: some nineteenth-century cities with their hundreds of thousands of smoldering soft-coal grates coughed amid a richer and deadlier smog than any modern city can concoct. But while in some ways air pollution is not as bad as it used to be, it threatens to get a lot worse than it is unless society revises the traditional orientation of engineering toward efficiency and economy.

The best antipollution intentions come up against certain gritty realities. The hunger for energy seems insatiable. Combustion of fossil fuels—particularly coal and petroleum products, which are the worst polluters—is likely to increase for quite some years to come, probably reaching several times present levels by the end of the century. The only courses left open, then, appear to be cleaning up combustion and substituting noncombustive modes of energy production. Whichever course society chooses—or even if it chooses to do nothing—men will pay a higher price for their energy.

Byproducts of combustion make up roughly 85 percent of the total tonnage burden of air pollutants in the U.S. Most of the emissions commonly classified as pollutants are not inherent in combustion, but are rather the results of inefficient burning or impure fuels. The pollutants fall into two main types, particles and gases. The particles include fly ash, which is an unburnable mineral fraction of ordinary coal; soot, which is burnable but unburned carbon; and lead, the unburnable additive in gasoline. Less visible but more damaging are the gases. These include sulphur dioxide (SO_2), a product of the combustion of coal or oil that contains sulphur; carbon monoxide (CO), emitted when insufficient oxygen is present during combustion; and various oxides of nitrogen, products of very high combustion temperatures. A great many complex hydrocarbon compounds, both particulates and gases, also result from incomplete combustion.

Few would quarrel with the view that air pollution is aesthetically objectionable. Numerous studies confirm it to be an economic drain upon society as well. Government economists at the National Air Pollution Control Administration in North Carolina have taken several routes to

The worst air-pollution threat for the future appears to lie in the relentlessly growing demand for electric power. As shown on this graph, drawn from government and industry data, power production capacity in the U.S. may be headed for an almost eightfold increase during the next forty years—unless society's priorities and policies change. Nuclear energy, often hailed as the answer to utility-industry air pollution, will not overtake energy from combustion of coal until the late 1990's (even if economical breeder reactors are developed in the near future, as is assumed in these projections). By then, coal consumption by the utility industry in the U.S. will have tripled from present levels. Already exacting a serious toll upon health and property, the gaseous emissions from coal burning would blanket vast areas of the U.S. if not controlled. As of now, no full-scale facilities have been perfected to remove the worst power-plant emission, sulphur dioxide gas. Even with the most favorable assumptions about the introduction of nuclear power, federal air-pollution officials predict that by the year 2000 electric utilities will have to spend somewhere between $450 million and $1.4 billion a year at present-day prices—depending upon the level of control desired—to control sulphur dioxide emissions

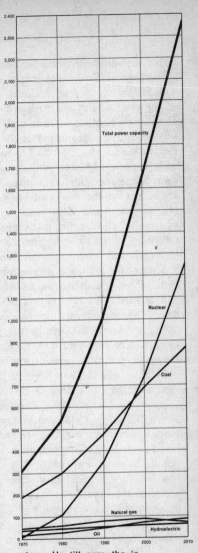

alone. Up till now, the industry's *total* outlay for controlling air pollution of all kinds has amounted to only about $1 billion.

arrive at admittedly broad-brush estimates that air pollution costs U.S. citizens somewhere between $14 billion and $18 billion a year in direct economic loss.

The potpourri of airborne particulates is the main cause of soiling—whether of shirt collars or of office buildings. But sulphur dioxide accounts for most of the damage to materials and much of that to agriculture. SO_2 combines with oxygen and then with moisture to form sulphuric acid. Sometimes this takes place in the lungs of animals, sometimes on the leaves of plants, sometimes in droplets of rainwater, and sometimes simply in the atmosphere where the acid persists as a fine, floating mist. The atmosphere of many industrialized areas is more corrosive to metals and other materials than sea air.

Under the influence of sunlight, several of the hydrocarbons react with the oxides of nitrogen to form ozone and a variety of complex organic compounds. Many of these "photochemical" substances are particularly damaging to plants. Because of the prevalent photochemical smog, leafy crops such as lettuce and spinach can no longer be grown in parts of southern California.

Sulphur dioxide has been implicated as the cause of many deaths in several air-pollution disasters, including those in the Meuse Valley in Belgium in 1930, in Donora, Pennsylvania, in 1948, and in London in 1952. These occurred when atmospheric temperature inversions combined with low wind ventilation to trap coal smoke over populated areas. Few doubt that any of the pollutants, not just SO_2, are harmful in concentration. The real question concerns the hazards of the long-term, low-level exposures encountered in everyday life. A great many studies strongly suggest that SO_2 is a cause or intensifying agent in various respiratory ailments even without inversion situations. The gas seems to do its worst in conjunction with particle pollutants, which can carry the SO_2 deep into the lungs and hold it there against sensitive tissue.

Similar evidence is also building up against nitrogen

dioxide as a contributing cause of respiratory ailments. Carbon monoxide, a well-known poison, has been measured at toxic levels in the streets of certain cities. Hydrocarbons and ozone are implicated in the eye-irritating smogs typical of Los Angeles and other sunny cities. Some animal experiments suggest that the hydrocarbons also cause lung cancer.

But the medical evidence against most air pollutants is far from conclusive. Even in the most polluted areas, citizens still breathe air whose contaminants are many times less concentrated than the pollutants in a lungful of tobacco smoke—and it took years to establish a tight case against cigarettes. After one surprising series of experiments, an independent laboratory under contract to the electric power industry has found that guinea pigs that have spent their life breathing an atmosphere with a fairly high concentration of SO_2 seem to live longer on the average than guinea pigs breathing pure air.

One combustion product that worries some scientists a great deal is not usually classified as a pollutant. This is carbon dioxide (CO_2), which unlike CO, SO_2, and the rest, is an inevitable result of carbon combustion. Hundreds of millions of tons of CO_2 are discharged into the atmosphere each year. Though CO_2 constitutes less than a tenth of 1 percent of the earth's atmosphere, the amount has increased roughly 25 percent during the past hundred years or so, and is expected to increase another 25 percent by the year 2000. CO_2 probably poses no direct threat to health, but quite a few scientists maintain that in the long run it may prove to be the most important pollutant of them all.

In theory, at least, an increase in atmospheric CO_2 should tend to raise the average worldwide temperature. The CO_2 acts like the glass in a greenhouse in trapping heat near the earth's surface. Accordingly, it is sometimes argued, the continued buildup of CO_2 will inescapably lead to calamitous overheating of the earth. Despite the theory, however, the average worldwide temperature has actually registered a slight drop since 1940. To account for this puzzling phenomenon several meteorologists are pointing suspiciously at another combustion product, suspended particulate matter, or smoke. Careful measurements indicate that the global atmospheric "turbidity"— or murkiness—has indeed increased over the past century, and thus could have offset the CO_2's "greenhouse effect." The particles 'not only shade the earth, but also act as "condensation nuclei" to promote cloud formation and so further reduce the amount of solar energy reaching the earth. Some meteorologists have even begun to worry in advance about the widespread operation of supersonic transport planes, because their release of condensation nuclei at high altitudes might greatly increase the cirrus cloud cover over much of the earth.

Whether the climate gets warmer or cooler, the implications are serious. Man and his institutions everywhere are critically adjusted to just the climatological conditions that prevail. Remarkably little is known about the interactions between earth and atmosphere, but it may well be that the nature of our environment is such that relatively small perturbations could trigger latent instabilities. Taken individually, the various inflictions that man imposes upon his environment may be tolerable; but in combination, they could work to disastrous effect. For instance, the expected exponential increase in worldwide combustion could combine with man's widespread destruction of the vegetation that removes CO_2 from the atmosphere. This, in turn, could lead to a world warming trend. Since water's capacity to absorb CO_2 decreases as the water gets warmer, the result might be still more CO_2 in the atmosphere and further warming from greenhouse effect, and so on.

In terms of sheer tonnage, the automobile is the prime U.S. air polluter, contributing about 40 percent of the 200 million tons of emissions that human activities put into the air in a year. The wide range of demands upon automotive engines—both cold and hot starting, high and low speeds, rapid acceleration, idling—entails compromising both combustion efficiency and fuel purity. Share of nationwide tonnage, to be sure, does not tell all the story. Given enough time and space to do her job, nature can cleanse the air of almost anything. Pollution is still very much the special bane of places where the time and space are lacking—where there is high traffic density, for example, or where people live too close to smokestacks. Much of the automobile pollution is discharged in open country where nature has a chance to work. Moreover, the auto emits little SO_2.

While efforts to find a substitute for the internal-combustion engine have captured much recent attention, probably the most productive approach to the problem of auto pollution for a couple of decades at least will prove to be what the auto companies have been maintaining—cleaning up the present internal combustion engine or its fuel, or both. Despite intensive and expensive effort, no one yet appears to have come close to devising a marketable alternative propulsion system that can do more than nibble at the edges of the national problem. No prospects are in evidence for inexpensive, lightweight batteries or fuel cells that could supply the combined range, speed, and hill-climbing ability that Americans will probably continue to demand. Several planners have pointed out another objection to battery propulsion: if most Americans were to drive electric autos, the power for charging the batteries would doubtlessly have to be drawn from the electric-utility system. This would mean something like a doubling of electric-power capacity and immense additional air pollution from power plants.

Neither the steam engine nor the gas turbine looks like a panacea at present either. It is probably no fluke that the internal combustion engine won out during the evolution of the automobile. Despite centuries of engineering attention, steam systems are still heavy and complicated in comparison with internal-combustion engines. Because they operate under more uniform conditions, gas turbines emit less carbon monoxide and hydrocarbons than piston engines per pound of fuel. But gas turbines burn more fuel per mile, especially in city traffic.

Even if alternative modes of propulsion do emerge, the sheer economic, political, and social inertia represented by all the institutions that have grown up to feed and care for conventional automobiles would seem to preclude a sudden shift. Happily, the near-term problems posed by the internal-combustion engine appear to be somewhat less pressing now than they did a year or so ago. Federal auto-emission standards, which were first applied to 1968 models and will become increasingly more stringent through at least 1975, have already begun to have discernible effects. The inevitable CO_2 aside, federal authorities estimate that auto pollutant tonnages, which totaled about 80 million tons per year in 1968, will be reduced to around 75 million tons this year and will continue to decline. If new exhaust standards that the government is pushing are adopted, the emissions should drop to around 55 million tons per year in 1985. After that, however, with continued increases in the number of cars and car-miles, the levels may begin to climb again. Adoption of some of the contemplated standards will hinge upon the development of a workable device —such as an afterburner—that would complete the combustion of unburned components of exhaust.

For the near term, at least, this process of modification will be less costly in either money or performance than alternatives to conventional internal-combustion engines. Conceivably, however, with sufficient attention to development, a respectable alternative could emerge. One possible approach is a "hybrid" vehicle with electric propulsion and a small combustion engine used to charge the batteries, but only in open country. Because this engine would turn at constant speeds and be equipped with exhaust afterburners, it would emit only a tiny fraction of the pollutants now put forth by the conventional engine.

The most nagging problems of pollution and pollution-control policy for the future appear destined to spring from the electric-power industry. Though electric-power generation now produces only 13 percent of the total pollutant tonnage, it accounts for more than 50 percent of the sulphur dioxide, about 27 percent of the oxides of nitrogen, and nearly 30 percent of the particulates. As the chart on page 117 shows, present trends indicate that soft-coal consumption by electric utilities will increase by a factor of roughly three and a half between now and the year 2000. This projection reflects anticipated demand for electric power as well as the great economic inertia provided by long-life coal-burning plants that are already built or committed. Nuclear energy's role will rapidly increase, but its rate of growth will be restrained by limitations in manufacturing capacity and by practical policies, in the electric-power industry and in government, of maintaining both a competitive alternative to nuclear power and a viable coal-mining industry—a large, labor-intensive activity that cannot be turned on and off with ease. While implying no inevitability, the projection for coal points to the magnitude of the air-cleaning tasks ahead. If sulphur dioxide and oxides of nitrogen are of concern now when electric utilities emit some 25 million tons of these pollutants a year, how much worse will the problem be when they burn three and a half times as much coal?

Since coal disappeared years ago from the cellars of most houses, Americans have come to think of it as old-fashioned. But coal still provides most of the energy propelling the U.S. economy. While estimated reserves of oil, gas, and uranium are reckoned in terms of a few decades, coal reserves are apparently ample for centuries.

The roughly $1 billion the electric-utility industry has spent so far for controlling air pollution has gone mostly into filters and electrostatic precipitators for removing particulate matter, and partly into premium prices paid for fuels with low sulphur content, including desulphurized oil and natural gas. For the future control of gaseous emissions most people in the industry would prefer to rely upon these measures, plus tall smokestacks. Tall stacks, industry spokesmen say, would permit the emissions from even very large plants to be adequately diluted before reaching the ground. One appealing quality of tall stacks is that compared to alternatives they are cheap, costing considerably less than other control devices to put in and nothing to operate. But outside the industry there is sharp disagreement about the effectiveness of tall stacks, especially as more and more generating plants dot the landscape. Great Britain has been equipping power plants with tall stacks, and one apparent consequence is that Scandinavian countries across the North Sea are occasionally pelted by rains and snows with a high content of sulphuric acid.

For control of sulphur dioxide, federal authorities would prefer removal of sulphur before the fuel is burned, or removal of sulphur dioxide from the flue gas. Both oil and certain kinds of coal can be desulphurized at an additional cost of about 10 percent. Already a great deal of desulphurized oil is being sold in cities, New York for example, has imposed stringent standards on SO_2 emissions. Potentially, at least, SO_2 could be removed from stack gases through any of a variety of approaches. Each of these approaches is likely to be expensive, adding somewhere between 10 and 20 percent to the cost of generating power, compared to about 1 percent for the tall-stack approach. The actual penalty for sulphur removal depends upon a number of variables difficult to quantify. In terms of capital outlay, the cheapest method involves injection of limestone into the combustion flame, where it combines with oxides of sulphur to form a solid, calcium sulphate. Calcium sulphate in copious quantities —1,100 tons per day from, say, a 1,000-megawatt power plant—would present an expensive waste-disposal problem, and, if not carefully managed, a potential water-pollution problem as well.

Other methods of SO_2 removal involve higher capital and operating costs but yield as a byproduct either sulphur or sulphuric acid that might be sold to offset some of their cost. By the 1990's, however, power generation might produce these chemicals so abundantly as to wreck the market for them. In that case, they too would become costly liabilities. Even now, while U.S. sulphur producers are busy extracting some 12 million tons a year, fossil-fuel burners are putting more than 16 million tons into the atmosphere.

Imposition of stringent and expensive controls on pollutants from combustion of coal or oil would no doubt lead to broader use of alternative fuels, and would also hasten the emergence of alternatives to the combustion process itself. One substitute fuel is natural gas, which has little sulphur content and in burning emits smaller quantities of oxides of nitrogen and hydrocarbons than coal, oil, or gasoline. Some urban generating plants now operate part of the time on gas. A number of enthusiasts envision a large future for compressed natural gas as a motor fuel. But natural gas is the least abundant of the fossil fuels, and this fact alone will limit its role largely to residential heating.

Proper economic inducements—including stricter emission standards—would also make the gasification of coal attractive. Relatively pollution-free, coal gas would necessarily cost a lot more than its thermal equivalent in solid coal, but its use as a fuel might prove cheaper than the other technological approaches to cleaning up coal combustion.

Noncombustive means of generating electricity have a long history; at one time, indeed, the term "generator" called up in the public mind an image of dams and water turbines. But hydroelectric power, which amounts to about 12 percent of total installed capacity in the U.S., is already approaching full exploitation here, as in most other industrialized nations. A number of other clean and noncombustive technologies, including geothermal power, tidal power, and solar power, are under development, but their potential uses appear to be limited to certain regions or special applications.

Fuel cells can produce electricity directly from a variety of substances. The use of hydrocarbon fuels, such as natural gas, entails—as always—emission of carbon dioxide. But other fuel-cell reactions do not even produce CO_2. With pure hydrogen and oxygen, the byproduct is nothing more objectionable than water. Other possible fuels are hydrazine and reformed ammonia, both compounds of nitrogen and hydrogen, which react with oxygen to produce electricity plus nitrogen and water. Though potentially the cost of some of these fuels could approach that of coal, fuel cells will be far too expensive for widespread use until the cost—especially of their catalysts—can be brought down. Costs of catalysts now range around $2,400 per kilowatt of generating capacity. This compares with *total* costs of around $100 per kilowatt of capacity for steam-power plants.

Ultimately, it seems inevitable that nuclear reactions will largely supplant combustion reactions in the production of electric power, especially in the very large central generating stations that are increasingly the fashion. Right now, though, nuclear energy is being slowed by some fundamental problems, aside from high cost and slow deliveries. One particular difficulty is the fear that leads citizens to prefer a fossil-fuel plant as a neighbor to a nuclear plant. A lot of concern has been voiced about radiation, including the possibility of such substances as iodine 131 finding their way into food chains and thence into the bodies of human beings. But actual measurements seem to show that the radioactive substances normally emitted by nuclear plants are virtually undetectable in the surrounding environment. Some nuclear plants actually emit less radioactivity than many fossil plants. Assuming that a continuing increase in electric-power demand is inevitable and a choice must be made, the evidence available suggests that nuclear plants pose less threat to the environment and to human well-being than fossil plants.

A valid objection to nuclear power is the thermal pollution that present-day plants pour into rivers and streams. To prevent their fuel elements from melting, nuclear plants produce steam at a lower temperature than fossil-fuel plants. This means that for an equivalent amount of electricity nuclear plants produce more steam. After passing through the turbines, the steam must be rapidly condensed; otherwise the power plant could not operate efficiently. To accomplish this condensation, the power plant draws large volumes of cooling water from a stream, pumps it through a condenser, and then returns it to the stream. The warmed-up water discharged by a nuclear plant is at about the same temperature as that from a fossil-fuel power plant, but there is about 40 percent more of it. By 1980, nuclear plants alone will require about one-eighth of the total volume of stream flow in the U.S. for cooling. Possible remedies include paying a penalty in generating efficiency by operating with less coolant volume; discharging the heat far out at sea; channeling the warmed water to cooling towers that discharge heat into the atmosphere; or, finally, putting the heat to use for space heating or industrial purposes. A bonus of the last approach is that it would reduce the burden of pollutants from combustive activity. By the mid-1980's the thermal problem should be moderated by the introduction of other kinds of reactors, notably breeder reactors. Breeders can operate with high-temperature fuel elements. The result is higher thermal efficiency and less discharge of waste heat.

As things stand now, there is reason for concern about the adequacy of the reserves of cheap nuclear fuel. Reasonably assured reserves of uranium in the U.S. would be insufficient to last beyond the 1980's if nuclear technology were limited to present types of reactors. More uranium will no doubt be discovered, but there is no way of knowing how much, or what it will cost to find and mine. Worries about fuel, however, should vanish for practical purposes with the development of the breeder reactor, which will create more fissionable fuel than it consumes. Even so, the fast breeder under development now is a difficult and skittish device, the safe operation of which may require expensive safeguards and so keep the price of electricity high.

In the past few months, scientists have gained renewed encouragement over the prospects for controlled nuclear fusion. The new optimism comes from experimental evidence—mostly obtained by Russian scientists—apparently indicating that the long-standing technical barriers to fusion are soluble, at least in principle. If and when they are overcome—five, fifteen, or thirty years from now—man will then have available an inexhaustible, cheap, clean, and safe source of energy. That could spell the end of man's grand-scale need for his ancient friend and enemy, combustion.

But the widespread deployment of fusion reactors is a long way off—perhaps half a century or more. Meanwhile, unless all the people of the world adopt new and unlikely attitudes that radically deflate the value of energy as an index of human advancement, we are likely to see fossil-fuel consumption continuing in its recent trend, at least doubling every twenty years. Even at present, some 15,000 cubic miles of air are fed into the flames of combustion each year, emerging greatly depleted in oxygen and poisoned as well. The time has come when the earth's atmosphere can no longer be regarded as limitless, but must be regarded rather as an exhaustible resource.

The traditional goals in engineering have been maximum performance and minimum cost in the objects being engineered. Now the objects being engineered must often include the very biosphere itself, the earth's thin, life-supporting peel of rock, soil, water, and air. All costs must be refigured, and many will increase. The costs may be paid in the form of nonproductive capital or operations, in compromised performance, and sometimes perhaps in simple self-denial. But if the payments are not made early, mankind will end up paying far greater penalties later—to property, to peace of mind, to health, or to life itself.

VIII
It's Time to Turn Down All That Noise

by John M. Mecklin

In the Bronx borough of New York City four boys were at play, shouting and racing in and out of an apartment building. Suddenly, from a second-floor window, came the crack of a pistol. One of the boys sprawled dead on the pavement. The victim happened to be Roy Innis Jr., thirteen, son of a prominent Negro leader, but there was no political implication in the tragedy. The killer, also a Negro, confessed to police that he was a nightworker who had lost control of himself because the noise from the boys prevented him from sleeping.

The incident was an extreme but valid example of a grim, and worsening, human problem. In communities all over the world, the daily harassment of needless noise provokes unknown millions to the verge of violence or emotional breakdown. There is growing evidence that it contributes to such physical ailments as heart trouble. Noise has become a scourge of our land, a form of environmental pollution no less dangerous and degrading than the poisons we dump into our air and water; it is one of the main causes of the exodus from our cities.

In many ways, noise is the most difficult form of pollution to combat. It has been recognized only recently as a major evil. It works so subtly on the human mind that it has gained a form of acceptance. Shouting over the din of an air compressor in New York, a newsman asked a construction foreman what he thought about the noise problem. "What are you," the foreman shouted back, "some kind of a Communist?"

Noise pollution is hardly a new evil. The word itself derives from the same Latin root as nausea. Noise bothered Julius Caesar so much that he banned chariot driving at night. In 1851, Arthur Schopenhauer wrote about the "disgraceful . . . truly infernal" cracking of whips in German streets. In a study published in October, 1955, FORTUNE reported "a rising tide of noise [in] U.S. streets, factories, homes, and skies" and asserted that Americans "have decided that noise should be abated." The optimism was unwarranted. Today the level of everyday noise to which the average urban American is exposed is more than *twice* what it was in 1955, and the cacophony continues to mount: the crash of jackhammers, whirring air conditioners, snarling lawn mowers, family arguments penetrating the paper-thin walls of homes and apartment houses, and the blast of traffic on freeways. Everyday noise is assaulting American ears at an intensity approaching the level of permanent hearing damage, if indeed the danger point has not already been passed.

As often happens with a slowly encroaching evil, it has taken a major outrage to stir up public concern. This is being provided spectacularly by the jet aircraft now bombarding some 20 million Americans every few minutes with a thunderous roar around our major airports. In the area of New York's John F. Kennedy Airport alone about one million people (including the students in about ninety schools) live within a zone of "unacceptable annoyance," as an official of the Federal Aviation Administration describes it. At Shea Stadium baseball games, the racket regularly drowns out not only the national anthem but also the players calling for fly balls. In Washington, jet noise so disrupted a ceremony attended by President Johnson at the Lincoln Memorial in 1967 that he ordered an aide to call the airport and stop it. In Los Angeles, concerts at the Hollywood Bowl have become virtually inaudible, and residents of nearby Inglewood have filed lawsuits that could total as much as $3 billion against the city.

The federal government is beginning belatedly to recognize that it has a problem—an awakening that may be partly due to the fact that approach paths to Washington National Airport pass directly over the homes of numerous top government officials. Congress voted to give the Federal Aviation Administration authority for the first time to fix aircraft noise limits. Though the effort may result in higher fares for air travelers, there is reason to hope that jet noise may be rolled back almost to the level of propeller planes within a few years.

But the same cannot be said for noise pollution in general. With a few exceptions, the steps taken so far have been palliatives. Memphis, for example, has enforced a ban on automobile horn blowing (instigated by an angry newspaper editor) since the 1930's. In New York in 1948 a landmark court decision upheld for the first time an award of compensation to an industrial worker who had suffered gradual hearing loss without losing work time. This type of claim is now recognized in some thirty states. Studies indicate, however, that only a small percentage of workers with legitimate claims have gone to court.

In Washington, at least a dozen federal agencies have become involved in the noise problem, and there has been one action of significance. That was the promulgation of a series of new regulations last spring under the Walsh-Healey Public Contracts Act of 1938, which limit industrial noise levels in most plants doing business with the government. The new regulations were initiated by the Johnson Administration, however, and were watered down by the Nixon Administration. Elsewhere a handful of unofficial organizations are agitating for noise abatement, notably in New York City.

Multitudes of special interests are arrayed against anti-noise measures whenever they are contemplated. When a state law was proposed to ban playing transistor radios in public vehicles, Buffalo radio-TV station WGR editorialized against such "inanities." Of about 125 industry representatives who testified at Labor Department hearings leading to the new industrial noise standards, more than 90 percent were opposed to regulation. Sample argument: "It is unrealistic and literally impossible to comply with." The cause of noise abatement wasn't helped any when the *Journal of the American Medical Association* argued in an editorial that "some noise must be tolerated as an unavoidable concomitant of the blessings of civilization."

To permit this kind of thinking to prevail is the true inanity. Virtually all man-made noise, whatever its source, can be suppressed. While some major problems, such as thin apartment walls and the roar of New York City's subway, would cost large sums of money to correct, many of the most irritating noises could be reduced at negligible cost. The screech of truck tires on pavement, for example, can be reduced at no extra cost or efficiency loss by redesigning the tread, and a quiet home lawn mower costs only about $15 more than the usual ear-jarring model. Some other examples of added costs: a garbage truck $2,400 (on top of an original cost of $15,600), a small air compressor $500 ($5,300), and on most machinery an additional 5 percent atop the original cost. In some cases there is also a relatively small cost in reduced efficiency. Mass production of silenced equipment would lower costs still more.

The expense becomes even less formidable when measured against the savings from noise suppression. The World Health Organization estimates that industrial noise alone costs the U.S. today more than $4 billion annually—in accidents, absenteeism, inefficiency, and compensation claims. The human costs in sleepless nights, family squabbles, and mental illness are beyond measure, but they surely must be enormous.

In the cases of air and water pollution, one of the main obstacles to corrective action is the large governmental outlay required—e.g., for nonpolluting municipal incinerators and sewage-treatment plants. But society's noisemakers are predominantly privately owned machines, many of which wear out and must be replaced within a few years anyway. Moreover, noise is not a uniquely big-city problem of little interest to suburban or rural taxpayers who are not exposed to it. Modern technology, its root cause, is everywhere, from grinding dishwashing machines in farm kitchens to outlying airports and thundering throughways that can be heard for miles.

The first and perhaps the most important course of action is to generate all possible public pressure on governments. It is no coincidence that one of the world's most effective anti-noise programs emerged in West Germany after the leading political parties there began including it in their election platforms. Once it becomes clear to Americans that noise is not an inescapable fact of life, that something *can* be done about it, and at manageable cost, the support for real action could be overwhelming. Says Judge Theodore Kupferman, a former Congressman from Manhattan and long-time anti-noise crusader: "In addition to the merits of the anti-noise cause, I don't see why more politicians don't take up the cudgel. Who's going to be in favor of noise?"

The most effective approach to governmental action probably lies in *federal* regulation. The legal authority already exists in the laws providing federal regulation of interstate commerce, health protection, and such specific functions as federal guarantees of housing loans, and there is a precedent for federal regulation of noise in the 1965 legislation empowering the Department of Health, Education, and Welfare to set limits on air pollutants emitted by motor vehicles. At the same time, state and local governmental action against noise, as well as support from enlightened businessmen, could go a long way toward reducing the problem, and perhaps in setting a trend—as a few localities have already demonstrated.

Noise—commonly defined as "unwanted sound"—works on humans in two ways. One, of course, is to cause deafness through deterioration of the microscopic hair cells that transmit sound from the ear to the brain. A single very loud blast, as from a cannon, can destroy the cells by the thousands and they never recover. (The Veterans Administration is spending about $8 million a year on the claims of some 5,000 servicemen whose hearing has been damaged by gunfire in training or combat.) Constant exposure to noises commonplace in our society can cause slower deterioration as the hair cells gradually rupture. There is a glimpse of the remarkable "redundancy" of the human body—in this case in spare hair cells—in the fact that all of us in modern communities have lost a substantial portion of our hearing mechanism without ever missing it. An experiment conducted by Dr. Samuel Rosen, a leading Manhattan otologist, has shown that aborigines living in the stillness of isolated African villages can easily hear each other talking in low conversational tones at distances as great as 100 yards, and that their hearing acuity diminishes little with age.

The second effect of noise upon humans is psychological and intensely personal. It relates not only to a lifetime of experience, but also to mood. Thus the scream of a siren at night may bring fright and anger to a thousand neighbors, but it means hope to a desperate accident victim. The human ear, unlike the eye, has no lids and cannot be turned off, not even in sleep. Nature's initial purpose in providing animals with hearing presumably was to alert them to enemies, with its function in communication coming at a later stage. Thus the instinctive human reaction to noise, especially unexpected noise, is fear and an impulse to flee. Children play games with this "startle effect," as psychologists call it. But to older people, just home from a hard day's work, for example, a sudden noise like the slam of a door or an automobile backfire or even the bell of an ice-cream vendor often can tip the balance of self-control and lead to an emotional eruption. Studies of sleep patterns have shown that people never "get used" to noise; on the contrary, the annoyance, and loss of sleep, worsen as the interruptions persist. This is the main reason why there is, rightly, such strong opposition to the use over populated areas of supersonic airliners whose sonic booms would cause psychic havoc among millions.

Clinical evidence has established conclusively that excessive exposure to noise constricts the arteries, increases the heartbeat, and dilates the pupils of the eye. Sigmund Freud wrote that noise could create an anxiety neurosis "undoubtedly explicable on the basis of the close inborn connection between auditory impressions and fright." One French study goes so far as to suggest that noise is the cause of 70 percent of the neuroses in the Paris area, compared with only 50 percent four years ago, and it blames noise for three recent premeditated murders. John M. Handley, a New York authority on industrial acoustics, wrote that "symptoms of hypertension, vertigo, hallucination, paranoia and, on occasion, suicidal and homicidal impulses, have been blamed on excessive noise . . . 'Noise pollution' may be one of the reasons why the incidence of heart disease and mental illness is so high in the United States." Other authorities have suggested that noise may be related to stomach ulcers, allergies, enuresis (involuntary urination), spinal meningitis, excessive cholesterol in the arteries, indigestion, loss of equilibrium, and impaired vision.

Tests of the effects of noise upon animals have produced dramatic results. Prolonged exposure has made rats lose their fertility, turn homosexual, and eat their young. If the noise is continued still longer, it eventually kills them through heart failure. There is clearly a limit to the amount of noise that any animal, including humans, can tolerate. But at what point does the noise in our daily lives begin to be dangerous? Dr. Vern O. Knudsen, former chancellor of the University of California at Los Angeles and a leading authority on noise pollution, believes we have already reached it. "Noise, like smog," he says, "is a slow agent of death."

There is no single universally accepted criterion of what constitutes excessive noise. The most common noise yardstick is the decibel (db) scale, which is an expression of the sound pressure that moves the ear. The scale begins at zero db, which is the weakest sound that can be picked up by the healthy ear. Thereafter, because of physical laws, the scale increases as the square of the change. Thus so soft a sound as human breathing is about ten times greater than zero db, while an artillery blast is one thousand trillion (1,000,000,000,000,000) times greater. To simplify things, the scale is in logarithmic form so that ten times the minimum is 10 db and one thousand trillion times the minimum is 150 db. The db scale does not, however, take account of the tones in the sound being registered—i.e., the frequencies of the sound waves being propagated. Scientists' attempts over the years to work out techniques to weight such factors for accurate registration of the ways that noise sounds to humans have led to a plethora of measuring scales.

There is agreement that high-pitched tones are more annoying and thus should be given more weight than low tones, but there the agreement ends. The most common weighting system is the "A" scale, written dbA, which gives less weight to low tones and thus more nearly matches the effect of sound on people. Beyond that the variations are myriad. For example: dbC ("C" scale), PNdb ("perceived noise"), EPNdb ("effective perceived noise"), SIL ("speech interference level"), and the "sone" and "phon" scales.

Some sample noise readings in the dbA scale at distances at which people are commonly exposed:

```
Rustling leaves ........................ 20 dbA
Window air conditioner ................. 55
Conversational speech ................. 60
(Beginning of hearing damage if prolonged ... 85)
Heavy city traffic ..................... 90
Home lawn mower ...................... 98
150-cubic-foot air compressor ........... 100
Jet airliner (500 feet overhead) ........... 115
(Human pain threshold ................ 120)
```

In the case of the laboratory rats mentioned earlier, death occurred after prolonged exposure at 150 dbA, or the equivalent of continuous artillery fire at close range. The take-off blast of the Saturn V moon rocket, measured at the launching pad, is about 180 dbA.

Very roughly, the noise level in busy sections of American communities is doubling every ten years. It has reached the point today where it is often greater than industrial noise levels. The main cause of the trend is the constant growth of the use of power. To cite one of the main new offenders, air conditioners are now in use in some 32 million homes. The giant machines on the top of large buildings often spew out more than 100 dbA and bother people for blocks around. No fewer than 89 million cars (up to 70 dbA) and about 18 million trucks and buses (up to 95 dbA) are cluttering our roads and streets. Millions of them are operating with defective mufflers, which always wear out faster than the vehicle. The beauty of our winters has been defiled by the din of some 700,000 snowmobiles, and our buses, trains, parks, and streets by millions of transistor radios. The racket in a modern American kitchen rises as high as 90 dbA midst an ever expanding profusion of dishwashers, mixers, grinders, exhaust fans, disposers, and the like. The National Institute of Mental Health is considering a proposal to wire up a typical housewife with telemetry like a spacecraft to try to study the effects of the pandemonium on her nervous system.

On top of all that, architects, engineers, and contractors in the $90-billion U.S. construction industry behave, says one acoustical expert, "as though they were born without ears." Thousands of new apartment buildings and homes are being thrown together like cardboard dollhouses, creating multimillion-dollar "noise slums," as one occupant puts it. Privacy, so badly needed by city dwellers, vanishes among the sounds of flushing toilets, electric razors, and family intimacies penetrating the walls, inhibiting conversation, and worsening tensions. Air-conditioning, heating, and ventilation ducts are made smaller and smaller to save space and weight, with the result that machinery must be faster, and therefore noisier, to move the same amount of air. Outside, meanwhile, some three million construction workers all over the U.S. create daily bedlam with jack-hammers, air compressors, earth-moving equipment, riveters, and similar mechanical monsters.

Americans often seem to react to noise as if it were a narcotic, as though nature were compelling us to accept it, even savor it, rather than engage in a hopeless struggle. Researchers have found, for example, that workers in noisy jobs often refuse to wear ear plugs because they are proud of their ability to "take it." In truth, this kind of tough-guy syndrome seems to be a subconscious device for sublimating discomfort. Psychologists think a similar narcotic effect may help explain why teen-agers sit for hours in rock joints, overwhelmed by "music" (as high as 130 dbA) that blots out all else in the world and, like marijuana, enables them to escape temporarily from reality.

The first real test of the nation's capability to roll back the engulfing tide of noise will be the FAA regulations on aircraft. The FAA will begin by fixing noise limits on the new generation of planes such as the huge 325-passenger Boeing 747. The ruling probably will stipulate that such planes must generate the equivalent of no more than about 95 dbA at a point about four miles beyond the start of the take-off roll; today's big planes register as much as 105 dbA. An improvement of that magnitude is being built into the 747's by Boeing engineers.

To supplement the new regulations, the National Aeronautics and Space Administration has launched a $50-million program to subsidize development of still another generation of even quieter engines through design and engineering innovations to slow engine fan-blade tips below supersonic speeds and thereby lessen the noise-making air turbulence. The agency hopes this will permit it a few years hence to begin further reducing the limit for new planes, perhaps to the equivalent of heavy city-traffic noise.

A third move will apply noise limits to the 2,000 airliners now in service on the nation's airways. The problem is much tougher in this case and no decisions have yet been reached. There is a good chance, however, that there will be a dramatic program, to cost about $2 billion, to "retrofit" the whole airline fleet with engine silencers. How such a program would be financed remains to be worked out, but it seems likely that the government would require passengers to share in the added costs; a general fare increase of 5 percent has been suggested. A "retrofit" program might be accompanied by a change in flying procedures. Instead of a gradual three-degree approach to airports, aircraft would make a two-segment approach, first at six degrees, then at three degrees for the final segment. That would permit them to stay longer at higher, and thus quieter, altitudes than is possible today. Such changes in procedure would also involve design changes in the aircraft.

In the area of general noise pollution other than aircraft, the one step that Washington has taken—new regulations under the Walsh-Healey Act—has been a disappointment to anti-noise advocates. The regulations benefit some 27 million workers in about 70,000 plants, but exclude millions of others in plants with fewer than twenty workers and less than $10,000 in government contracts, thus omitting small businesses where abuses are no less deplorable.

The Johnson Administration, which initiated the action, originally proposed to fix a noise limit of 85 dbA, with higher levels permitted for short periods. The proposal was so hotly opposed, however, especially by high-noise industries like textiles, that the Nixon Administration compromised on a maximum of 90 dbA—or 5 dbA more than the experts regard as safe. Even at 90 dbA, however, the new regulations will have a notable, indeed historic, impact if they are enforced. At least half of American industry today permits noise levels above 90 dbA. The American Petroleum Institute estimates the cost of compliance to the oil industry alone at $40 million to $50 million to modify its existing equipment.

Meanwhile, the federal government is acquiring a great deal of valuable expertise in studies ranging from apartment-house noise insulation to a computerized analysis of transportation noise. By far the most significant of the studies is an exhaustive document called "Noise—Sound Without Value," published last fall by a special ten-agency committee. The report asserted: "Increasing severity of the noise problem in our environment has reached a level of national importance and public concern." With notable political courage, it added that the solution "frequently will require actions that transcend political boundaries within the nation," i.e., it should not be left to the states. Not long afterward, the Johnson Administration, which had promoted the study, went out of business.

Which points directly to a central unknown today: what will the Nixon Administration do about noise? The top authority is the newly created Environmental Quality Council, headed by Nixon himself, which to date has made no decisions on noise. There is fear among anti-noise advocates that Nixon's strong feelings about state responsibility may lead him to stay out of noise control except perhaps for voluntary—and therefore ineffective—guidelines. Such concern is one reason why the Senate passed a bill proposed by Washington's Democratic Senator Henry Jackson to create a prestigious, independent council to recommend policy on all forms of environmental control, including noise. A similar bill is pending in the House.

If the Administration should leave non-aircraft noise pollution to the states, the outlook could be gloomy; few states have taken actions of any consequence. California has a law limiting the vehicle noise on freeways to 88 dbA, and a noise abatement commission will begin hearings aimed at producing recommendations by 1971. The law is so loosely enforced, however, that a Los Angeles police official confessed he did not know it existed. In New York State, indignant citizens along the roaring New England Thruway, where some 10,000 trucks create a steady din around the clock, persuaded the state legislature to fix a limit of 88 dbA on each vehicle. There have been only sixty-three arrests since 1965—and the maximum fine is only $10 anyway. Connecticut also plans to introduce a noise-abatement program.

In the long-suffering core cities, where noise works its greatest evil, the record is spotty. Several cities have anti-noise ordinances, e.g., Dayton, Dallas, Chicago, and Minneapolis. In San Francisco the Bay Area Rapid Transit System now under construction is spending $1,250,000 (only one tenth of 1 percent of the total cost) on noise suppression; it is expected to be the quietest subway in the country—85 dbA on the platforms versus New York's numbing average of 102 dbA. Milwaukee attempted to reduce truck noise by a city ordinance, only to have it overturned by the courts on grounds that it invaded state jurisdiction; the effort was laughed into obscurity anyway when a newsman discovered that the city's own vehicles were violating the ordinance.

Despite its multitude of other problems, New York City probably has tried harder than any other big community to mount a really effective anti-noise campaign. Much of the initiative came from a group of volunteers called Citizens for a Quieter City, Inc., headed by Robert Alex Baron, forty-nine, who was so incensed by the din of a construction project outside his Manhattan apartment that he quit his career as a Broadway play manager in 1966 to do something about it. His efforts helped coax Mayor John V. Lindsay to appoint, in 1967, a special anti-noise task force of technical experts and public-spirited citizens. These and other pressures combined to persuade the city council in 1968 to pass the first building code of any major U.S. city with an anti-noise provision. It requires that new residential buildings must be constructed to cut noise penetration by about 45 dbA, which is appreciably less strict than the codes of several European countries but nevertheless a major stride forward.

To prove that Americans do not have to live with noise, Baron arranged for a public demonstration of silenced machinery in New York's Lincoln Center one day in 1967. Among other items, he displayed a quiet air compressor imported from Britain. Within a few months, at least one major American manufacturer, Ingersoll-Rand, began actively promoting a similar machine. The city is running a test of the feasibility of using paper bags for garbage instead of cans, and it has contracted with General Motors to develop quiet garbage trucks to replace the present fleet of "mechanized cockroaches," as Baron calls them.

In a report to be published this month, the New York task force has also recommended an extraordinarily ambitious program for further steps, ranging from new zoning rules to creation of a corps of noise inspectors, with the objective of reducing the noise level in busy areas to 85 dbA and the residential level to 40 dbA in daytime and 30 dbA at night. The New York initiative is attracting considerable local attention. On N.B.C.'s Johnny Carson TV show, Baron appealed for people annoyed by noise to write him; the result was some 2,500 letters, most of them venting long-pent-up anger and frustration.

But the effort has a very long way to go. The anti-noise forces to date have failed even to persuade the New York Police Department to try to enforce the ordinance against needless horn blowing. The problem is compounded by the fact that the city has no direct authority to act against noisy vehicle engines (which are the state's responsibility), or noisy aircraft (the FAA's), or even the New York subway system (the Transit Authority's).

There is no mystery about how to control noise. At least sixteen European countries have building codes with anti-noise provisions, many of them tougher than anything even contemplated in the U.S. The Soviet Union, which began a "struggle against noise" in 1960, says it has banned factory noise above 85 dbA and limited the level in residential areas to 30 dbA. The West Germans, among other actions, have set up no fewer than eight categories of noise limits; they offer tax concessions and easy credit to manufacturers willing to silence machinery acquired before the limits were established; and they stamp the maximum noise permitted each vehicle on the owner's driver's license. In France, needless horn blowing has been successfully outlawed—much to the surprise of the French themselves—and other noise regulations are so well enforced that a peasant recently was fined $50 for a noisy cowbell.

Unlike other forms of pollution, noise comes from an infinite number of sources and cannot be cut off simply by cleaning up a few big operations such as garbage dumps. The answer, however, does not lie in brave new proclamations. Ways must be found to get at the problem through appeals to the self interest of business and community leaders and through governmental regulations that are realistic and easily policed.

Two further federal moves are needed now to provide a legal framework for minimum national standards. One is to broaden the recent anti-noise regulations to protect workers in all factories engaged in interstate commerce. The second is to invoke the interstate commerce principle to permit the fixing of limitations on the noise created by the machines that industry produces. The objective of such a move would be to oblige manufacturers to design noise suppression right into their goods, and its national application would guarantee that no company would be hurt competitively. Says Leo L. Beranek of Cambridge, Massachusetts, chief scientist for the nation's largest acoustical consulting firm: "We have got to have noise regulation at the federal level. Controls in only a few scattered cities won't work; quiet products must have a national market."

A proposal for federal action on such a broad scale obviously invites innumerable problems. For one thing, it would require congressional action. It would compound the infighting already under way among the several federal agencies competing for anti-noise responsibility. Whatever the bureaucratic machinery, however, the approach probably should be to seek a broad mandate from Congress, and then to begin application of specific controls on a progressive basis, beginning with the most urgent problems, such as highway vehicle noise and outrageously noisy machines like air compressors.

Federal laws, of course, are no panacea. At best they can be expected only to provide minimum standards that could then be reinforced through state and local action. In some cases such action can be carried out most easily through local regulation—e.g., anti-noise insulation of residential buildings, which can be enforced with relative ease through existing building inspectors. The mere existence of federal anti-noise laws would create strong psychological pressures on local governments and industry to act.

Apart from legislation, there are innumerable other opportunities for action. The federal government buys something like 35,000 vehicles annually. To require such vehicles, especially trucks, to be fitted with good mufflers, quiet tire treads, and other noise-suppressing equipment would go a long way toward encouraging manufacturers to make such items standard equipment. State and local governments could easily do the same. Procurement policy can be similarly useful across a broad spectrum of other items that are purchased by both consumers and government agencies—e.g., garbage cans, which can be quieted for about $1.50 apiece. The Federal Housing Authority and other national and local agencies have the power now to make compliance with noise standards a condition for publicly backed loans. The National Park Service has the same kind of authority now to bar noisy vehicles, transistors, and the like from our national parks. It should also be feasible for the federal and local governments to grant tax concessions to encourage industry to suppress noise. By relatively simple fiddling with electronic circuits, buses and trains could be fitted with jamming devices to discourage transistor addicts. And automobiles could be equipped with two horns, one for highways and a quieter beeper for city streets, as is widely done in Europe. The Federal Highway Safety Bureau revealed that it is already considering a requirement of this sort.

Contrary to the view among some industrialists that noise control is an expensive luxury, it is in fact good business; the lack of effective noise control at the source, moreover, is bad for business. In the case of aircraft, anti-noise flight procedures—e.g., disuse of some runways—are further reducing the capacity of our congested airports, while popular reaction against aircraft noise is making it increasingly difficult to find sites for new airports. Cities like New York are losing tens of millions of dollars in traffic diverted elsewhere.

For industry in general, the mounting cost of hearing-loss compensation claims could easily become astronomic if workers began going to the courts in large numbers. In view of the growing evidence that noise is a significant health hazard, it would make eminent sense for insurance companies and labor unions to add their considerable weight to the battle. With a few exceptions, businessmen have been surprisingly slow to recognize that noise prevention can be marketed; for example, in advertising for quiet apartments or noiseless kitchen equipment. There is also a major public-relations consideration in the growing feeling among environmentalists that corporations have no more right to dump noise on communities than air and water pollutants.

"Let avoidable noise be avoided," said the late Pope Pius XII in a 1956 appeal from the Vatican. "Silence is beneficial not only to sanity, nervous equilibrium, and intellectual labor, but also helps man live a life that reaches to the depths and to the heights. . . . It is in silence that God's mysterious voice is best heard."

IX

Downtown
Is Looking Up

by *Walter McQuade*

The downtowns of most large American cities are sleek, bleak, monotonous, impersonal, and inconvenient, expressing an over-all meanness of environment in the midst of great wealth and almost unlimited resources of building materials and architectural talent. After business hours they are as lifeless as unplugged computers. And they occupy the most precious land on the continent: a square foot of central Houston costs up to $110; of Atlanta, $100; of Chicago, $500; of San Francisco, $250; on Wall Street, $600, or $27 million an acre.

The icily aggressive central business district—or CBD in the jargon of city planners—is not that way through neglect. Of the 130 American cities over 100,000 in population, perhaps half have had their business districts worked over in the past twenty-five years by committees of bankers, builders, real-estate men, elected officials, architects, and midtown merchants. Their efforts have changed the American look downtown from a 1920-ish eclecticism to a glowering monotony. Gertrude Stein's devastating remark about Oakland, California, "There is no there there," no longer applies, as it did once, to the typical American CBD.

But in most cases the strong new "there there" has as

many drawbacks in human terms as the old. It is not just the architecture of individual office structures that is the villain, for there are occasional estimable buildings among the shelves of canned office space. And within, the individual office rooms are usually impeccable. They have superb air conditioning, soft fabric on the floor, floods of shadowless fluorescent illumination. But there is almost nowhere for the ordinary employee to go to eat a tolerable lunch except the company cafeteria because so many modest restaurants have been rooted out by the real-estate progress. There are more subjective lacks as well. In a report to the mayor of New York three years ago by a committee composed largely of corporate executives, the pleasures now disappearing from the Manhattan environment were identified as: diversity, coherence, grandeur, style, and humanness.

In a recent book, *American Architecture and Urbanism,* historian Vincent Scully of Yale pointed to thirty-seven-year-old Rockefeller Center as "one of the few surviving public spaces in America that look as if they were designed and used by people who knew what stable wealth was and were not ashamed to enjoy it. Flags snap, high heels tap: a little sex and aggression, the city's delights. Jefferson would have hated it." By contrast, he wrote, postwar urban developments "all seem to indicate in their various ways that Americans can no longer put the centers of cities together at all but can only destroy them." It all goes to suggest that in the twenty-five years of manic management expansion since World War II there has been precious little breadth added to the urban white-collar working environment.

What makes the polished blankness of the standard kind of downtown development seem most impoverished is the fact that a far broader approach is being taken in some cities of North America. For the first time since the early 1930's, when the late John D. Rockefeller Jr. sat at his desk quietly selling off Standard Oil of New York shares at $2 in order to pay for the ice-skating rink, the other

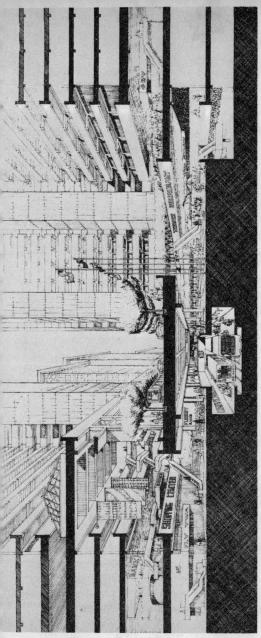

A busier but less congested Seattle downtown has been designed by architect Rai Okamoto of the San Francisco firm, Okamoto Liskamm. Included are most of the elements of the new-style central business district: direct connection to mass transit, diversion of automobile traffic, easy pedestrian circulation on several levels, inclusion of restaurants, stores, plazas, open spaces, planting, and other spatial pleasantries. But design is only one step in the lengthy, complicated process of building a new central business district. Seattle's voters turned down the subway proposal first time around. Planners predict the congestion will get worse and the voters will then change their minds.

amenities, and the over-all spatial wisdom of Rockefeller
Center, some of the people who shape CBD's are reaching
out beyond today's glassy-eyed aura of office efficiency,
beyond the easy kind of real-estate profits, for a degree of
genuine environmental congeniality. A few scattered monuments to this new movement have already been completed
in places as far separated as Montreal, Baltimore, Boston,
and Minneapolis. Others are under construction from
Dallas to Chicago and from San Francisco to New York.

These new developments do not lack diversity, coherence, or humanness. They have so much of these qualities
that they sometimes seem to be aping the street life of old
European towns. The employee on his way to the office in
the morning travels a convenient, even amiable, route. He
crosses a cobblestoned square as in Vienna, or a brick
courtyard as in Venice. Splashing fountains culminate
careful vistas. Thickets of little trees grow out of the rubble pavement, or stand on it in gigantic pots. Gallerias
modeled on Milan's compete with outdoor sitting areas.
Instead of conventional street lamps, there are wands
crowned with large settings of electrical jewelry. As in Bo-
logna, arcades shield pedestrians from rain. Cultivated little gardens condense nature into the mix; sculpture adds
scale. Batteries of convenience stores with a stylish bou-
tique touch are inserted between the inevitable banks and
brokerage offices. Future projects will include saunas and
tennis courts for lunchtime use. There are outdoor cafés,
an over-all concentration on creating a pedestrian street
life. The automobile is banished outward or downward
from the premises.

Other basics of the new midtown include particular emphasis on bringing back people of varied incomes to live
in the CBD—in subsidized projects, in private apartments,
in hotels, or even in town houses. Existing ethnic settle-
ments are treasured, not just tolerated. Legitimate theatres
are provided whenever troupes can be found to staff them
and audiences to support them. Movie houses are in the

mix, usually occurring in pairs of small art houses. To the outdoor cafés are added indoor restaurants of varying degrees of exoticism and price range, insistently more interesting for lunch than company cafeterias.

Even more important are the links to transportation. The skeletal characteristic of the new midtown is again reminiscent of Rockefeller Center, a closely cultivated access to mass transit. Subways are the spine, and are accessible under cover. In Chicago, for example, a new subway will be burrowed to connect all the parts of that city's spreading CBD, and most new office buildings will be able to tap into its service underground. Although the new subway is not yet there, the network of connections is already being formed. Because the new facility will principally serve the CBD, replacing and extending the famous loop of the elevated and carrying office workers to the commuting lines, much of it will be paid for by a special real-estate tax levied just on owners in that tax district. Planner Vincent Ponte, who worked out the superb sub-surface circulation system of midtown Montreal, has devised an underground circulation even broader for Dallas. The route of San Francisco's subway has also been recognized as a nucleus for further growth of the midtown office district, and underground connections from new office buildings are being calculated. In both Chicago and San Francisco, sunken plazas will admit daylight to some of the subway platforms.

At its best, this shrewd souping up of a very dull environment is bringing a genuine liveliness back to midtown. At first experience it may seem a little glib, too easy a copy of what charms us in Europe, but it is nothing less than a serious attempt to add complexity to the single-mindedness of so many new office districts by larding other aspects of metropolitan life into them. Not since the statuesque "city beautiful" movement of the turn of the century in America has there been such a turn-on of effort to do something about physical amenity in urban life. None of it aids directly in American cities' most severe physical problem —the condition of the slums, which have suffered from the same arid architectural approach generally applied in the CBD. But a CBD reveals a great deal about how a city sees itself, and it may not be too much to hope that a more civilized design approach downtown will stir echoes throughout town. What any kind of downtown development can contribute to the city is hard cash. The $2.3 billion of new construction in Chicago's business center over the last ten years has added an annual $84 million to the city's tax receipts.

The new midtown is a difficult real-estate package to wrap, because it is much more expensive than the conventional kind of business development and yet must pay for itself. It needs a chorus of supporters: business leaders with verve, vision, and tenacity, enlightened planners and architects, and city officials who are also eager to try something new and more difficult in the way of development. The old way of stacking office buildings along the standard old avenues, producing the standard congestion, is quicker and more profitable, at least in the short run. It has the added virtue of avoiding the political infighting made necessary by any reordering of the usual midtown real-estate priorities, now directed largely to maintaining the full values of existing blue-chip locations. What the new way requires, more precisely, is a larger assemblage of land, not just the usual parcels, and a great deal of extra time and trouble. "It is a young man's game," says architect I. M. Pei, who has been at it as long as any of the professionals. "Developers used to think they could do it in three and a half years. It takes at least twice that."

The fuel is enthusiasm. The top city planner of San Francisco, Allan B. Jacobs, admits that planning is not the principal ingredient: "If there is a concern with the quality of the environment on the part of the whole culture—the elected officials and the businessmen included—that is what counts." Frederick J. Close, board chairman of the Aluminum Co. of America, declares that conventional commercial success is not enough. "We are developing Century City [in Los Angeles] to make a profit. . . but seek to do so through a concept that will enrich the lives of those who work and live there . . . man cannot live by elevator shafts alone. He has to have an occasional view of a sunset, a cool morning breeze, and a mud puddle . . ."

Everyone agrees the single most vital man in the process is the local apostle of change, who is usually a businessman or a government leader. Baltimore architect Archibald Rogers says, "There always has to be the impresario, a Sol Hurok to sell the idea to the community, and to stick with it, giving it continuity." In Baltimore, the impresario was J. Jefferson Miller of the Hecht Co.; in Boston it was Mayor John F. Collins and Development Administrator Edward J. Logue; in Philadelphia it has been a series of businessmen plus city planner Edmund Bacon; in Dallas, businessman Mayor Erik Jonsson. Whoever they are, the impresarios need not only broad support, but a clear view of where they are headed.

The disease of most office districts has usually been identified as overcrowding; nevertheless, the essential idea of the new downtown is actually to *increase* the density of people working in concentrated areas. The main real-estate function downtown is the mass production of rental office space, and every additional million square feet of office space adds as much as $10 million in annual rentals, as well as about 5,000 employees to the sidewalks, most of them at about the same times of day. For comfort there has to be better circulation than conventional sidewalks, and better circulation usually demands large-scale planning, several blocks at a time. With all those people on hand, there are also some profits to be made in the rental of stores, etc. But the correct balance of components needs an expert hand to tilt the rent structure properly to encourage restaurants, apartments, and such notorious losers as legitimate theatres. Not only does all office space make for a very dull downtown, but good restaurants can't make it with just a lunch trade.

The most complete, competent, and exciting of the plans for new midtowns is the blueprint for Yerba Buena, a twenty-five-acre redevelopment project in San Francisco. The details of it illustrate how density can work, while demonstrating, too, the necessity for extensive government involvement in most downtown reconstruction. The plan was announced in June of last year by M. Justin Herman, executive director of the San Francisco Redevelopment Agency, and a man who qualifies for any list of official Huroks. Herman had already made his mark on the city by guiding the pioneering Golden Gateway Development of commercial space and housing.

Yerba Buena will include a convention center and a commercial complex on about three blocks just south of Market Street. On its many levels will be a 350,000-square-foot exhibit hall, a 14,000-seat sports arena, a 2,200-seat theatre, parking garages with 4,000 stalls, a downtown air terminal, an Italian cultural center, a number of office buildings and a hotel. There will be direct underground connection to the new Bay Area Rapid Transit system when that is completed. The team of designers includes not only such local design stars as architects Gerald McCue and John Bolles, landscape architect Lawrence Halprin and city planners Livingston and Blayney, but also architect Kenzo Tange of Japan.

Several developers are competing to undertake the $200-million-plus enterprise, including four American groups and Impresit, a subsidiary of Fiat of Italy. In it for the developer is the prospect of a large, valuable site ready for use, offered at a low price. In it for the city is the chance to upgrade an entire neighborhood and to have private money put up facilities needed to attract large conventions.

The key power in this sort of undertaking is the city's right to condemn land and to write down its cost with the aid of federal funds, under urban-renewal legislation first adopted by Congress in 1949. The Yerba Buena land carries a valuation of $12 a square foot, only a tenth as much as sites on the other side of Market Street. Golden Gateway was made possible by the same process, as was its new neighbor, Embarcadero Center, another artful elevated plaza to be connected by pedestrian bridges to the Gateway. On land costing only $31 a foot, the Embarcadero developers are building 2,851,000 square feet of office space, a hotel, a set of shops, a brace of theatres, various appropriate social interspaces, and a movie house. Although there appears little likelihood that more federal money will soon be available for downtown renewal, that key power to condemn land and resell it to private developers has been well established. Missouri, Illinois, Massachusetts, and New York now have their own laws to permit cities to assemble the sites and then wholesale the land to developers at cost, or even at a profit.

Very different from the Yerba Buena approach in method is a project called Crown Center now under way in Kansas City. It was undertaken from the start by Hallmark, the greeting-card company, and has no public money in it at all. Crown Center first began to form in the mind of Joyce C. Hall, president—and principal owner —of Hallmark, in 1955 when the company completed a headquarters building twelve blocks from the downtown Kansas City core. Hall continued to buy land in the area until he held the better part of a hundred-acre zone edged by parks and hospitals. When he became chairman of Hallmark, and his son, Donald J. Hall, succeeded to the presidency, the younger Hall brought in architect Edward L. Barnes of New York to master-plan the development and design office space, plazas, and pedestrian ways. Harry Weese of Chicago was invited to design a hotel, and Norman Fletcher of the Architects Collaborative in Cambridge, Massachusetts, to do apartments.

An immense cavern for a concealed parking garage eventually to take 2,500 cars is being carved out of the Missouri limestone in the middle of the site, and the frame of the first office building is complete. When the project is finished, it will have 1,100,000 square feet of office space, 2,200 apartments, a coterie of special shops, a pair of cinemas, and restaurants with a continental touch. Hall has run the same kind of market research he uses to determine what the public will buy in greeting cards (Hallmark sells eight million of them a day) to find out what level of sophistication it will support in specialty shops, and in victuals.

There are those in Kansas City who question the basic need for a $135-million complex a dozen blocks removed from the present city business core. Hall points out that when the project is completed it will have only 11 percent of the office space in the city and 4 percent of the apartments. But there is no doubt that Hall is creating a new focal point downtown, against which the old CBD will be judged. No outside financing has yet been arranged for the enterprise, although Hallmark has more than $13 million of equity in it already.

Some of the very same doubts mentioned about Crown Center were current about Rockefeller Center in New York forty years ago. But in the forty-year interval, not only has that project prospered as a demonstration of excellent over-all planning of an office district, but ordinances in numerous cities have been tipped in the direction of such development.

In the past, most municipal manipulation of density had taken the form of zoning ordinances, which were essentially lists of limitations on the height and bulk of buildings and thus primarily negative. Some cities went so far as to limit the number of stories a building could extend up above street level. Developers in San Francisco could build up to only sixteen stories, in Chicago to only forty stories, in Pittsburgh to only ten stories.

Today, however, most of these height limits have been removed, and the maximum extension upward of the building is usually determined by a calculation based on the size of the lot being occupied. If, for example, a developer owns a plot of 50,000 square feet in a district in which the floor-area ratio permitted is ten, he can put up a building containing no more than ten times 50,000 square feet, or a half-million square feet net. If the building occupied the entire plot it would be roughly ten stories high. Or he could put up a more slender tower twenty stories high, with each floor containing 25,000 square feet, leaving half the ground floor space open.

To move zoning further off the negative, cities have begun to govern CBD development in two new ways. Large projects, involving several square blocks and located near mass transit, are treated as a single site. Zoning restrictions are suspended, in exchange for the city's right to govern the over-all balance of buildings, circulation space, and other amenities. The city itself sometimes prescribes the details of its CBD as Boston did with the Government Center project, replacing old Scollay Square with a group of city, state, federal, and privately owned office buildings. Even as the bulldozers went to work on Scollay Square, architect I. M. Pei was engaged to decide which streets in the razed area should be retained and which eliminated, what kinds of walkways and open plazas and subway connections should be provided, and even what size and height buildings should be built.

The space sculpture was enacted into law by the city council, and in the ensuing years has resulted in a handsomely balanced assortment of office buildings that seem to be generating still another boom. In 1961, before the beginning of the Boston downtown renaissance, the Boston Redevelopment Authority hired economist Robert Gladstone to estimate the need for new rental office space in that city for the succeeding fifteen years, and he came up with a figure of 5,500,000 square feet. The local real-estate fraternity was so skeptical that they tried to get the figure revised downward before it was published. But today, with five years of Gladstone's prediction period still to run, 22 million square feet of new office space are already in place or under way in old Boston.

Not all developments can be so sweeping, and so another kind of system has been evolved by cities for bartering densities for the kind of design that makes greater densities work. This is called incentive zoning. Under it, city planning commissions can permit private developers of office buildings to build in more square feet than the basic ordinance permits in exchange for including such conveniences as concourse connections to subway stations and sheltered pedestrian streets. Providing the extra amenities usually costs the developer extra money, but it is not difficult to figure out how much extra rental space he deserves to balance the added capital investment.

San Francisco is the city furthest forward in codifying its system of incentive bonuses; there are some ten, including one provision to permit an office-building investor to include an extra quantity of rental space in exchange for providing an observation deck on a high floor open to the general public. In New York it is incentive zoning that is being used to prevent the demise of the theatre district. A large office building on the site of the old Astor Hotel on Times Square will house a new $5-million legitimate theatre, an underground movie house, an arcade of shops, and three restaurants, including a café on Shubert Alley. The developers, Sam Minskoff & Sons and a Lehman Brothers group, say they do not expect to make much profit on those extra amenities. The carrot that tempted them was permission from the city to add two floors of valuable office space to their big tower. Three more legitimate theatres are under construction in other office buildings rising in the district; an additional five are in negotiation between the city planners and developers.

The Lindsay administration is pushing the program for economic reasons, not just out of affection for the arts. The attempt is to keep the oncoming glacier of office buildings from sweeping away the entire theatre district west of Sixth Avenue. The chairman of the Planning Commission, Donald H. Elliott, points out that the theatres are a core attraction of New York's fourth biggest industry, tourism, and that no new ones have been added to the area since the 1920's. Another use of incentive zoning is illustrated in the Wall Street area, where the new U.S. Steel building, now under construction, has been permitted to vary the usual bulk requirements in return for a public park on its own land across the street. Other such openings in the financial district will be sought under the official New York City Lower Manhattan Plan, whose principal authors were Wallace, McHarg, Roberts & Todd of Philadelphia.

Sometimes, instead of offering carrots, cities wield the stick of the many police powers they possess over building design within their borders. In Los Angeles, planning director Calvin Hamilton proposes to link new downtown buildings by means of a second-story walkway, thus lifting pedestrian traffic from the street level. The cost of extending these elevated pedestrian ways will be borne by private developers but street bridges will require public funds. The twin Atlantic Richfield office towers, fifty-two stories high, will be linked to their garage in the Bunker Hill redevelopment district across Fifth and Flower streets by such an elevated walkway. Eventually this upper-level walkway will extend to a new hotel and the Union Bank. In Dallas the idea is to ask all developers to install connecting underground concourses within their own plots, but to have government pay for the links that have to cross under streets.

Good mass transit is almost essential to developing an efficient new downtown district of real density; in its absence solutions are awkward. The voters of both Los Angeles and Seattle in recent years have killed proposals for mass transit, although both cities are burgeoning downtown, particularly Los Angeles. Within the once moribund inner freeway loop of that motorized city, over five million square feet of new office space have sprung up since 1960, as well as the first tower apartments ever built in the CBD, and much more is on the way. As an alternative to more expressways downtown, the city proposes to install islands of multilevel parking facilities perhaps five miles outside the business center itself, and to load the motorists into minibuses. Within a mile or two of the core, elevated moving sidewalks would take pedestrians to the elevated walkways within the core itself.

Some city governments are big enough builders in their own downtowns to set an example to private developers in the provision of amenities. When Chicago's towering new Civic Center was put up in the Loop, a sweeping public square for pedestrians was left open in front of it, a space that sets off not only the Civic Center itself, but all the other new and old buildings around it. One reason it is judged a great success is the fact that it has recently been attracting protest groups, as well as art shows and more placid manifestations of civic vitality. Now the First National Bank of Chicago is placing the same kind of wide-open plaza beside its new skyscraper two blocks away, and there will be a third open space beside the new Federal Building. These three considerable cavities in the density of the Loop provide Chicago with the kind of decompression that squares in old Italian cities offer, complementing each other. In contrast, the New York pattern of small plazas lined up jaggedly beside each other along an avenue add more confusion than relief.

The most common mistake in efforts to shape the new downtown is the same as in all real-estate dealings, an error in timing. William Zeckendorf was one of the first Sol Huroks to embrace the possibilities of the new kind of downtown, but his ebullience led him to underestimate badly the period necessary to ripen the real estate. Zeckendorf began in Denver in 1953, in Montreal in 1955, and in Washington, D.C., in 1956, but could not last it out financially. Other developers harvested what he seeded. Cities too can misjudge timing. Cleveland had a part of its downtown replanned by I. M. Pei & Partners in 1960, and undertook an immense clearance of existing buildings. But, as it turned out, the city's capacity to market the land to developers was overestimated, and some of the cleared land remains just that.

Another promotional mistake is to make the answers seem too easy, or to present a final design too early. Even a relatively weak pressure group in the community is usually powerful enough to ·fight down a pat solution. The classic case of the too complete solution was architect Victor Gruen's masterful plan for downtown Fort Worth, which was commissioned in 1956 by J. B. Thomas. It never got into the ground because the whole community had not been involved in its evolution. A more widely based Fort Worth group has recently commissioned Lawrence Halprin & Associates to have a go at solving the city's continuing downtown problems.

Archibald Rogers holds that there are four essential steps, almost a parliamentary procedure, for the consultant implementing CBD action. He admits that his firm, RTKL Inc., has become as much an architect of process as of design. First, says Rogers, comes the reconnaissance stage, in which the planners not only find out what the urgent problems are downtown but define "the personality of the milieu." In Cincinnati, for example, in 1963, a large part of this personality was represented by the three political factions in the city council, led by Charles Taft of the Charter party, Democrat John Gilligan, and Republican Eugene Ruehlmann. The reigning feeling of all three factions about downtown was frustration because three successive plans had been evolved, each to be blocked in turn.

When the council called in Rogers and his partners to try to come up with an effective plan, the architects were advised to avoid two actions that the council believed had helped kill the other plans. The public, the council was convinced, was dead set against the expense of a subterranean parking garage downtown. The other sore point would be any move to relocate an anachronistic old sculptural fountain, done in the Renaissance style and sitting in the middle of a street where people could look at it while stalled in traffic.

According to Rogers, the second step, after reconnaissance, must always be to reach a firm definition of objectives and strategies. This was done in Cincinnati by holding many meetings with civic groups and the public, by involving any public figure who could be interested first-hand, and by using the local press and TV for further stimulation. Rogers recalls, "After the objectives had been worked out, the council's review committee said, 'Now start drawing.' We said, 'No, the objectives haven't been ratified yet.'" The council did that by adopting a set of 250 ordinances.

Next came design—not just one design, but all the ways the architect planners could think of to meet the objectives ratified by the council. Says George Kostritsky, one of Rogers' partners: "If you don't present all the alternatives in design to everyone in the community, they will develop their own alternatives." Among the many tentative alternatives presented were moving the cherished Renaissance fountain and putting some parking underground.

The next to last step outlined by Rogers is presentation of the plan, which never should come as a real surprise to the community. And finally, after the plan has been approved, an immediate start must be made in construction to keep the document from lying on the shelf. "If the process is broken, what you have done immediately becomes obsolete," says Kostritsky.

It took six years in Cincinnati from the beginning of the process to the completion of the first component, quite rapid for downtown redevelopment. Moreover, on dedication day most of the Cincinnati citizens seemed more than pleased by a new pedestrian plaza in the heart of the downtown business district, with its scarlet oaks, locust, and linden trees, and its chess tables built in among the benches. It is called Fountain Square because the old fountain was moved in, with new plumbing inserted to make it somewhat more of a gusher. Under Fountain Square is also a municipal garage of 600 spaces that helps support the plaza structurally and, through its charges, financially as well.

Presented with all the other alternatives, the community had indicated this was what it wanted. Jefferson, a master of parliamentary procedures as well as an inventive architectural designer himself, might have approved, at that.

X

Conservationists at the Barricades

by Jeremy Main

Noisy, militant, litigious, growing in strength and numbers, the conservationists are on the march. They believe their mission is desperately urgent—that unless Americans change some of their attitudes toward the environment, the country will literally destroy itself. This may seem like an extreme notion, but clearly a growing number of Americans think the conservationists—or environmentalists, to give them a label that suits their newly expanded goals—may be right. The power of this burgeoning movement is being felt in the courts, in politics, and in the boardrooms of the nation's top corporations.

The number of conservation organizations and their size are growing too fast to tabulate. There are more than 150 national organizations and thousands of local groups. Two of the oldest organizations, the Sierra Club and the National Audubon Society, have doubled their memberships in the last three or four years. A local disaster can spawn overnight a vociferous group such as GOO, which stands for Get Oil Out (of the Santa Barbara channel). On Long Island, some young lawyers and scientists have set up the Environmental Defense Fund to pool their expertise in taking environmental cases to court. A new group called Friends of the Earth deliberately chose not to seek tax-free status so as to be able to lobby for new environmental laws and elect sympathetic politicians.

When the issue is urgent, the conservationists win allies in unlikely places. For example, the Citizens' Crusade for Clean Water, which encouraged Congress to vote $800 million for pollution control instead of the $214 million requested by President Nixon, is a coalition of some forty organizations coordinated by the Izaak Walton League. In addition to the Sierra Club and other predictable conservation groups, the cause attracted the A.F.L.-C.I.O., the United Automobile Workers, the U.S. Conference of Mayors, the League of Women Voters, and the National Rifle Association. Such alliances give the conservation movement a political weight that it has never had before.

Alarm over the environment has created a kind of "new politics" that transcends traditional party or ideological leanings. Even members of the John Birch Society and the S.D.S. can agree that clean air and water are desirable. The movement is drawing its strength from the grass roots as few others have done, because it grows out of the individual annoyance, discomfort, or even suffering of millions of persons.

Most conservationists begin as ordinary people upset by small grievances. The involvement of Mrs. Ellen Stern Harris, a plump, persistent Beverly Hills mother of two, began when she decided one day to complain that the palm trees on her street were not being trimmed as they were on other streets. After making periodic calls to city hall without any results, she suggested to the authorities she might get up a neighborhood petition. The next day the trees were trimmed. Mrs. Harris discovered she could change things. Today she is a member of the Los Angeles Regional Water Quality Control Board and the Los Angeles *Times* named her a Woman of the Year. She is a real force in goading indifferent southern Californians to care for their unique environment.

Students are the newest recruits to the cause. They were rarely seen at conservation meetings in the past, but now they are pouring in to join the movement. A conversation with a college or high-school youth today seems to turn inevitably to the environment. Environmental groups are sprouting on campus just as S.D.S. chapters did a while back. Students are parading, picketing, threatening boycotts of polluters, and turning up at conservation conferences to pepper the speakers with difficult questions.

The old "birders" and hikers are still the spearhead of the conservation movement. These old organizations, however, are bringing their goals up to date to fight a broader "environmental" battle. The Sierra Club, fighting fifty-five different legal actions around the country, even obtained injunctions against five members of the Nixon Cabinet to hold up projects the club did not like, such as Disney's Mineral King resort in the Sierra Mountains.

The Sierra Club wants as much as it did when it was founded in the 1890's to keep the wilderness unspoiled, and its agenda still includes traditional objectives such as saving the redwoods and finishing out the national park system. But a list of priorities drawn up by the club puts at the top a new item, "environmental survival," which encompasses air and water pollution, pesticides, and even urban planning and population control.

The National Audubon Society, which put together a coalition that stopped construction of a jetport in the Florida Everglades, is also expanding its goals. Gene Setzer, chairman of the board of Audubon, says, "We were established in 1905 for the specific purpose of saving an endangered species, the egret, from plume hunters in Florida. It is still our purpose to save an endangered species, only now the endangered species is man himself."

Such a conclusion has led the conservationists to challenge some basic ideas about life in a nation of plenty. Many conservationists believe that unless America abandons the notion that a growing population can prosper only through a growing output of goods drawn from an endless frontier of resources, and unless technological progress is restrained, then all the efforts to improve the environment will be in vain. Population control is the conservationists' ultimate goal.

As their objectives become more ambitious, the conservationists are adopting new, more militant tactics. The laws enacted to protect forests and parks or to clean up the air and water are not enough for them. As conservationists see it, the government agencies whose decisions affect the environment were created in times when there was little concern for the environment. These agencies shun new responsibilities and become afflicted with a kind of tunnel vision, which tends to focus less on the public interest than on promoting the industry the agency is supposed to regulate. To shake bureaucrats—and businessmen—out of their frozen attitudes, conservationists are deliberately seeking stormy confrontations, sometimes with celebrated opponents chosen in order to generate the maximum publicity. Later, when some big battles have been won, they may be more accommodating. Right now they want to establish their legitimacy and power.

The front line of this war is in the courts. Judges are more receptive to change than bureaucrats, and their decisions tend to have more weight and clarity. Joseph L. Sax, a Michigan University professor of law who is writing a book about environmental law, says, "We are beginning to see value in maintaining resources rather than merely exploiting them. The courts are going to have to respond to this new perspective."

The articulation of even basic principles of conservation poses a new challenge to the legal profession. At a conference on law and the environment held at Airlie House in Virginia, James E. Krier, acting professor of law at U.C.L.A., defined the problem: "The promised surge of environmental litigation calls for rethinking much of our substantive and procedural law. Much of that law was made during the prime of the old, proprietary lawsuit, which it suited well enough; it fits poorly, however, the frame of the new lawsuit brought to protect environmental (not economic) values in the public (not private) interest. The common-law concepts of nuisance and waste, for example, are not responsive to the needs of environmental litigation . . . they reflect a far too narrow and myopic view. . ."

Over the years many an environmental case has been won as a private damage suit—for example, when chemicals damaged crops or pollutants killed fish. The law can cope easily with monetary claims by an injured party. But now the conservationists are suing, not for themselves, but in the public interest, and not for monetary damage, but for damage to beauty, history, peace and quiet, wildlife, and trees and plants. To do this effectively they must win new precedents in court.

The strength of the legal approach in saving the environment was demonstrated in a marathon struggle between Consolidated Edison, the New York utility, and the Scenic Hudson Preservation Conference. Scenic Hudson, as the case is called, began in 1964 when a group of Hudson Valley property owners and conservationists challenged Con Ed's plans to build a two-million-kilowatt pumped-storage plant at a scenic and historic area on the Hudson River. Construction has yet to begin, and New Yorkers are still short of power. Over the years the case has been to three hearings before a trial examiner of the Federal Power Commission, and has gone up to the U.S. Court of Appeals for the Second Circuit, and Con Ed tried unsuccessfully to take it to the U.S. Supreme Court.

The essential decision in Scenic Hudson was delivered in 1965 by the Second Circuit Court of Appeals. It gave the conservationists two important precedents that have already been used in subsequent cases. First, the court ruled that conservationists have "standing" to intervene in hearings by the Federal Power Commission and in court. The FPC wanted to keep them out of the hearings on the grounds that the "aggrieved" parties entitled to intervene under the Federal Power Act include only those whose personal economic interests would be damaged by a power project. But the court ruled that the phrase aggrieved parties included those "who by their activities and conduct have exhibited a special interest" in "the aesthetic, conservational, and recreational aspects of power development."

The circuit court's second important ruling was that the FPC cannot "act as an umpire blandly calling balls and strikes for adversaries appearing before it" but has an affirmative duty to protect the public interest and to gather information on its own. In this case, the court found the FPC had not given enough study to the environmental effects of the Storm King plant or the alternatives to it. This decision means that small, underfinanced conservationist groups do not have to carry the whole burden of challenging all the reports and expert witnesses of the large utilities.

A second dispute about the Hudson River has placed another important weapon in the hands of the conservationists. Governor Nelson Rockefeller wants to build a six-lane expressway along the east bank of the Hudson, part of it extending on a river causeway bordered by a four-mile dike. A group consisting of an ad hoc Citizens' Committee for the Hudson Valley, the village of Tarrytown, and the Sierra Club sued the Army Corps of Engineers, the Federal Secretary of Transportation, and the New York State Commissioner of Transportation last January in a set of four cases. After an exhaustive five-week trial in the district court for the Southern District of New York, the group won a permanent injunction against construction of the expressway.

The conservationists, represented by David Sive, an overworked New York lawyer who is also chairman of the Sierra Club's Atlantic Chapter, won on several points. The court decided that the Corps of Engineers erred in not obtaining the permission of Congress to build the dike and causeway. The plaintiffs proved that these were indeed the kind of structures envisioned in the Rivers and Harbors Act of 1899, which specified that they may not be built in navigable waters without congressional approval. The court reaffirmed and extended the Scenic Hudson ruling that conservationists had standing in court.

The most significant decision in the expressway cases involved the subtle question of whether the actions of government agencies can be reviewed. Some agencies, such as the FPC, have been set up by laws that specifically provide for court review. But where there is no statutory provision, as in the case with the Corps of Engineers, the government argues that its "sovereign immunity" puts its actions beyond court jurisdiction. In the expressway cases, the court upheld the general right of review and jurisdiction over the actions of the Corps of Engineers and the New York State Department of Transportation. In an appeal to the Second Circuit Court of Appeals, the government argues that if the district court's decision becomes precedent, then almost anybody could be a plaintiff against government action on nearly any matter of general concern.

Young lawyers see in the conservation battle an opportunity to work in a higher cause, just as other young lawyers found a cause in the civil-rights movement. The Environmental Defense Fund, Inc., which operates from an attic room in the town center of Stony Brook, New York, has an influence quite out of scale with its modest offices and its $300,000-a-year budget. E.D.F. grew out of a suit in 1966 to stop Suffolk County from spraying public places with DDT. Although the suit was thrown out of court, the county stopped spraying. Since then, E.D.F., financed mostly through the National Audubon Society by donations from the Ford Foundation and others, has followed up with suits or petitions in Michigan and Wisconsin, and against the federal government. Roderick Cameron, the personable young director of E.D.F. who holds both engineering and law degrees and who began his career as a government lawyer, says his organization does not take on private damage suits because they solve only one case and pit one private interest against another. E.D.F. wants cases that will establish precedent and "incorporate modern science into public policy." The policy is to be legally daring but scientifically cautious, to reach for precedents in law with overwhelmingly strong scientific cases. To overcome the weight of expert witnesses that industry and government usually have on their side, E.D.F. has created a pool of 300 volunteer scientists, many of them top men in their fields, to testify in its cases.

The Washington partnership of Berlin, Roisman & Kessler represents another approach to environmental law. The firm occupies a suite on N Street facing the town houses occupied by the well-known partnership of Arnold & Porter. The similarities end there. When they were setting up shop, Edward Berlin and Anthony Roisman would take off their lawyers' clothes at the end of the day and put on dungarees to spend the evening painting and fixing up partitions, doors, and bookshelves. The two men, and Gladys Kessler, who joined them later, are all former government lawyers in their early thirties.

They hope one day to make a moderately comfortable living, but right now they are just covering their overhead working fifty to sixty hours a week on what they consider "public interest" law, which includes environment cases, for modest fees. Like E.D.F., they hope to break new ground by arguing, for instance, that the Ninth Amendment (which says that the enumeration of rights in the Constitution should not "deny or disparage others retained by the people") is the basis for a constitutional right to a decent environment. They are using the Fifth Amendment ("No person shall . . . be deprived of life, liberty, or property, without due process of law") to fight construction of an overhead power line on a property along the Potomac River in West Virginia. They maintain in this case that the court, to assure due process, should make sure there are no reasonable alternative routes for the power line.

It is a sign of the times that even "straight" law firms, as Berlin calls them, are now beginning to devote more of their time to unpaid *pro bono publico* work. Arnold & Porter has one partner working full-time on public-interest cases, including environment, and he can draw on up to

15 percent of the time of the other sixty-five partners and associates in the firm. They have helped conservationists in two cases by writing a brief and by preparing congressional testimony.

Through their suits the conservationists are beginning, slowly, not just to win cases but to reorient the system a little. They are achieving precedents that should change court and government procedures. After the decisions in the Scenic Hudson case, the FPC is not likely again to approve a power plant without investigating thoroughly the environmental consequences and the alternative sites. Such decisions can also influence legislatures. Congress passed an act saying there should be no more federally licensed or financed projects on the Hudson unless the Interior Department decides the project will not harm the river's resources.

The attack through the courts, however, has limitations. One is money. The Scenic Hudson case has cost the conservationists more than $500,000 (Con Edison's legal costs in the case are far higher).

The larger limitation is that many environmental problems cannot be given precise legal definitions. Although the courts may change government procedures, it is unlikely there will ever be an environmental code as definitive as the criminal code. Joseph Sax, professor of law at the University of Michigan, says, "We can state a principle which says people cannot embezzle funds from their employers. We can't deal with the environment in that way. It's a much more fluid situation. We don't want to say, for example, never fill a marsh, cut a tree, or dam a river. We are always looking for some subtle balance between industry and nature, between high-density and low-density uses, between preservation and development."

When it comes to evaluating the best use of land, even the conservationists can differ. Walt Disney Productions was picked by the U.S. Forest Service to build a $35-million resort at Mineral King, a valley on the threshold of the Sierra Nevada, 228 miles north of Los Angeles.

The state of California decided in effect to subsidize the venture by building a $20-million highway to Mineral King. But the Sierra Club obtained preliminary injunctions preventing the Secretaries of the Interior and of Agriculture from signing the agreements. The Sierra Club, which once elected the late Walt Disney to honorary life membership for "his magnificent contributions to a widespread appreciation of our wildlife," believes that America's "development psychosis" must be stopped. Says the club's executive director, Michael McCloskey, "The whole idea of parks is to provide nature sanctuaries that will be immune from the fads of the local people, that will last through the ages, that are not negotiable." Although the Disney organization has announced elaborate plans to protect the delicate ecology of the valley and has produced a panel of conservationists on its side, the Sierra Club fears the Mineral King project and its highway would be only the beginning of large-scale, uncontrolled development in the area.

Other conservationists in California disagree. They believe the value of a ski resort outweighs the value of leaving Mineral King in its pristine state. The California branch of the National Wildlife Federation supports Disney, arguing that Mineral King is already being ruined by uncontrolled visitors in the summer and that the project would be a good test of what private enterprise can do to balance the needs of recreation with the needs of nature.

The passionate conflict over the construction of a jetport in swamplands forty miles west of Miami is another illustration of the split between those who would preserve wilderness areas and advocates of economic development. Construction of the jetport was supported by the Dade County Port Authority, the Mayor of Miami, and the airlines. The opposition was organized principally by the Miami representative of the National Audubon Society, thirty-one-year-old Joe Browder, a former newsman and policeman.

The conservationists appealed to Washington, where

they won strong backing. The decision—accepted by state and local authorities—was that the Port Authority must find a new site and abandon the jetport. What the conservationists particularly feared was the pollution of the Everglades that might result from the commercial and residential growth that would accompany the development of the field.

For several years to come the conservationists' main tactic is likely to be confrontation rather than accommodation. This may stop many depredations, but it is essentially a negative force. If conservation groups want, they can delay construction of almost any project—a highway, a generating plant, or even a factory.

Russell Train, former president of the Conservation Foundation in Washington who was appointed Undersecretary of the Interior—reportedly to mollify those who objected to Hickel's appointment—is concerned there will be too much "environmental litigiousness." "We can't govern by protest, demonstration, and litigation," says Train. "This doesn't mean I am not delighted to see such energy channeled into this area. But you can't operate a going concern with protest as the main mechanism."

In private, some of the most intransigent among conservationists admit that the stress on obstruction is a temporary tactic. Eventually, they hope, the burden of defending the public interest in the environment will be taken up by public institutions such as new regional planning groups, and by existing agencies, such as the AEC and the FPC, if they adjust to environmental needs. Litigation runs counter to one of the most forceful arguments made by the conservationists: that the environment cannot be protected without comprehensive, long-range planning. Obviously a river cannot be cleaned up unless all the towns along the river get together and decide what to do. But nothing is likely to upset comprehensive planning more than endless litigation.

As concern over the environment grows, politicians respond. This political awakening is nationwide and spans the ideological spectrum. For a politician, it is the safest of all issues, since conservation has no admitted enemies. Massachusetts, the state that probably has the most advanced environmental legislation, also has a chief executive, Francis Sargent, who can say, "I'm the only governor who has come up the conservation route." He became state director of marine fisheries twenty years ago and spent most of the intervening time in one kind of state resources job or another. Sargent, a handsome, athletic man who owns a sporting-goods store on Cape Cod and wears a gold fishhook for a tiepin, decided to switch from appointive to elective office when he thought that concern with the environment had grown into a statewide issue. "I felt I had gone as far as I could outside elective office," says Sargent. "In elective office I could make things happen, rather than react to what was already done." He was elected lieutenant governor and then became governor when John Volpe was appointed Secretary of Transportation.

Among other things, Sargent moved to restrict the use of DDT (a day before the federal government announced its restrictions), participated in hearings aimed at getting new clean air standards, reorganized the state government, and created an office of environmental affairs in his cabinet. He now wants to set up an Environmental Quality Control Council, which will act as a sort of state ombudsman on environmental questions.

The climate for such legislation has been created largely by the Massachusetts Audubon Society, the most competent and one of the largest conservation groups in the country. "Mass. Audubon," as it is called to distinguish it from the separate national society, has a full-time staff of over 100. The director is Allen Morgan, forty-four, a businesslike bird watcher who at fourteen became the first non-Harvard student elected to the Harvard Ornithological Club. Mass. Audubon provides full-time ecology teachers to one-third of the school systems in Massachusetts. It gives lunches periodically at downtown clubs in Boston to tell businessmen about the problems of the environment.

The main function of Massachusetts' 280 conservation commissions established by state law is to try to turn local government into a friend rather than an enemy by making conservation a good financial deal. The commissions try to buy up urban land that might otherwise have been used for revenue-producing developments. Towns pay 25 percent of the cost of land, the state 25 percent, and the federal government 50 percent. In the Berkshires the commissions have the help of the Berkshire Natural Resources Council, which is financed by the Berkshire *Eagle,* General Electric, and other companies, and which helps find and negotiate acquisitions for the local commissions. George Wislocki, who runs the council, says he has "no long-range plan except the one I carry in my head. You buy land when it's on the market. The next ten years is going to decide the future of the Berkshires. It's a problem of land use. If we get nothing but quarter-acre lots for 'leisure homes' by the lake, and without sewers, we've had it."

Since conservation issues usually are local issues, the federal government has been slow to act. Senator Gaylord Nelson of Wisconsin, who makes conservation his principal field, says that in 1966 he could not get any cosponsor in Congress for a bill to control the use of DDT, but in 1969, when he reintroduced the bill, dozens of Senators and Congressmen wanted to be cosponsors. Spencer Smith of the Citizens Committee on Natural Resources, a conservation lobby, says that while six years ago he had to keep track of perhaps 250 bills during a session, last year he had 8,000.

During 1969 the Nixon Administration, preoccupied by Vietnam, appeared to be trailing the rest of the country in its concern for the environment. The President did little more than create a token Environmental Quality Control Council, consisting of Cabinet members who have other matters on their minds and rarely meet to discuss the environment. The Administration held down spending on the environment. Now, however, President Nixon is moving toward making the environment *the* issue of the 1970's.

Meanwhile, the conservationists are going through some

difficult adjustments. Traditionally, the movement has been supported mostly by professional, middle and upper class, urban WASPs on both coasts, more sociable companions of the trail than fighting reformers. They included a good share of misanthropes. "In every organization there may be members who dislike people as much as they like trees," says the Sierra Club's David Sive.

The new demands made on the conservationists, particularly the expensive lawsuits they have undertaken, are adding to their financial burdens. The Sierra Club was hit specially hard when the Internal Revenue Service canceled its tax-exempt status on the grounds that its efforts to stop the damming of the Grand Canyon were too political. That decision has lost the Sierra Club many a large donation. Other groups, fearful of losing their tax-exempt status too, have been careful to avoid any taint of political lobbying. David Brower seeks no tax exemption for his new organization, Friends of the Earth, and says it is frankly political. He intends to put his organization behind individual political candidates.

With their wilderness-for-the-WASPs background, the conservationists have trouble adjusting to today's broader environmental problems, which encompass the plight of the cities and the quality of life. Some clubs remain basically social. Even the Sierra Club's Los Angeles chapter rejected a black applicant as recently as 1959. Black faces are rarely seen at conservation meetings. But new groups are trying to step into the urban void. In Watts, there is an integrated group called the Urban Workshop, which operates under the motto "ghetto beautiful," and in Washington, D.C., blacks are working through a number of organizations, including one called Niggers Inc., to block superhighways. The conservationists are quite aware of the need to grapple with city problems, but they don't know how. David Sive, who grew up in New York City, says, "we haven't even begun to solve this problem."

XI

Victory on San Francisco Bay

by Judson Gooding

The tactics used in winning the Battle of San Francisco Bay will long be studied as a textbook example of how to wage a successful campaign in defense of the environment. Conservationists opposed to filling the bay found that the path to victory in the struggle to save natural resources leads from the mimeograph machine and the telephone through newspaper columns and radio stations to the halls of the legislature. There, after a campaign of an intensity unprecedented in California, they succeeded in bringing about passage of a bill that prohibits any indiscriminate bay filling. Henceforth only projects of priority importance to the community as a whole will be permitted. The work done by the conservationists to save the bay has a dual benefit for the rest of the country: it assures the preservation of a great national scenic asset, and it illustrates how citizen power can be mobilized to preserve other endangered wonders.

At stake had been the future of an immense, extraordinarily lovely series of inland seas stretching fifty miles from north to south, and extending inland to the great Sacramento delta where the waters from sixteen rivers flow down from the Sierras. The bay's opening to the sea, the Golden Gate, was carved out by the Sierra waters grinding their way through the coastal mountain range two hundred thousand years ago, but it was discovered only in the eighteenth century. An early explorer traveling with Juan Bautista de Anza wrote in 1776: "If it could be well settled, there would not be anything more beautiful." He described the harbor as "so remarkable and so spacious that in it may be established shipyards, docks and anything that may be wished." What man wished, it slowly became clear, was to fill the bay where he needed more shoreland, and to use its waters as a dump for garbage, sewage, and industrial wastes.

The first bay fill occurred in what is now downtown San Francisco, where ships that had brought prospectors in the 1850's gold rush were turned into wharfside houses and offices as the city grew explosively. A new type of fill operation began as hydraulic gold mining up the rivers washed hundreds of millions of tons of silt downstream to the bay floor, where it destroyed rich oyster beds and silted channels so that water-flow patterns were altered. At the northern and southern extremities of the bay, thousands of acres of bay and marshland were set off from the ebb and flow of the tides, for use as salt-evaporating ponds and as game preserves. By 1960 diking and filling had reduced the water area by one-third, from 680 to 400 square miles. A report by the Army Corps of Engineers showed that two-thirds of the bay was less than twelve feet deep, and thus "susceptible of reclamation," and that half of it belonged to private owners or local governments.

Complacency about the preservation of the bay was hard to understand, even before the engineers' alarming report was published. It was true that pollution of bay waters had been reduced markedly, so that swimming was once again possible in some areas. However, both San Francisco and Oakland were developing their waterfronts, and building up their big bayside international airports— relying on fill for new land. All around the bay, waterfront housing developments and marinas were being built. New refineries and other industries were under construction, or being planned, adding to the economic base of the thirty-two towns abutting the bay.

This rush of development threatened to destroy the bay. Studies showed that the bay-area population would double within thirty years, would reach thirteen million by the year 2020. Pressures to convert bay waters to valuable land were certain to increase enormously. Garbage dumping into diked-off areas became a commonplace solution to disposal dilemmas. During the late 1950's and early 1960's one city after another announced plans to expand its land areas into the bay, and private developers proposed numerous schemes of their own. In the south bay, Foster City was planned to accommodate 35,000 people on 2,600 acres, a scheme that required dredging 18 million yards of sand from the bottom of the bay.

One of the most startling proposals was advanced by the city of Berkeley, which planned to double its size by filling 2,000 acres of the bay. This plan alarmed many east bay residents, including Mrs. Clark Kerr, whose husband was then president of the University of California. She and two friends, Mrs. Donald McLaughlin, whose husband was chairman of the university regents, and Mrs. Charles Gulick, wife of a professor, investigated the Berkeley proposal. They quickly became convinced that conservation groups should be mobilized, and soon aroused the Sierra Club, the Save the Redwoods League, the Audubon Society, and others in defense of the bay. With this encouragement, Mrs. Kerr and her friends decided to form the "Save San Francisco Bay Association." They mailed out 700 letters and got 600 replies, many with contributions. This overwhelming consensus indicated that although concern over the bay had not been particularly visible, it was wide and deep.

The association persuaded the University of California's Institute of Governmental Studies to carry out a detailed study of the bay's future. The report, published in 1963, pointed out that fill reduced the amount of tidal flux and diminished the oxygenation of the water. The findings

helped to defeat the Berkeley Bay fill proposal after the
association made it a local election issue. The group then
broadened its attack and, working through Assemblyman
Nicholas Petris of Oakland, tried to get the state legislature to enact a moratorium on all bay fill. Petris (who
subsequently became a senator) offered two bills, but opponents crippled them with amendments providing for exceptions to the moratorium, and the bills died in committee.

Renewing the attack at the state level, the association
decided to convert a powerful politician to its cause instead
of working through an already convinced conservationist
with a narrow power base. Mrs. Kerr settled on Senator J.
Eugene McAteer, who was planning to run for mayor of
San Francisco. She urged him to head the fight for the bay
bill. "If you take the lead on this," she said, "you will be
'Mr. San Francisco' himself." McAteer threw himself into
the struggle with vigor. (He was to die of a heart attack before the fight was won.) He and Assemblyman Petris proposed legislation creating a Bay Conservation and Development Commission (BCDC) to look into ways in which San
Francisco Bay might be developed to its maximum useful-
ness without harming its scenic or recreation potential.
The study would also report on the effects of fill on the
bay. A crucially important provision prohibited any new
fill during the group's three-year study period without a
public hearing and authorization by the commission.

There was strong opposition to the bill in the legislature. But Save the Bay Association members flocked to
Sacramento, where they packed the legislative chambers,
and some Oakland activists even mailed small bags of sand
to their legislators with tags saying, "You'll wonder where
the water went, if you fill the bay with sediment." With this
kind of highly visible and audible support, McAteer was
able to get the bill passed.

The language of the McAteer-Petris act was direct. "The
bay is the most valuable single natural resource of an entire region," it said. "The present uncoordinated haphazard

manner in which San Francisco Bay is being filled threatens the bay itself and is therefore inimical to the welfare of both present and future residents of the area."

The law created a commission of twenty-seven members to be appointed from federal, state, and local agencies and governments, and the public. The commission was to present a plan for bay conservation and development to the legislature in 1969. If that plan was approved, it would be put into effect and the commission would become permanent. If the legislature demurred, the commission was to go out of existence—and there would be no curb on filling the bay.

The commission went to work on the plan, breaking down the study into twenty-five topics, such as recreation, water front housing, fish, and wildlife. The finished report, submitted in January, 1969, stressed that the bay should not be treated as "ordinary real estate," but as an asset belonging to the area, the state, and the nation, an asset to be protected for present and future generations. The commission held that since any change in the bay might affect other parts, this protection could be afforded only through a regional approach.

The report, bearing out earlier findings, declared flatly that "any filling is harmful to the bay" because it upsets the ecological balance, diminishes the bay's ability to assimilate pollution by reducing the water volume exchanged by the tides, and cuts down the surface area, thus limiting oxygenation. Commission jurisdiction over development of shoreline areas was one of the most controversial elements of the plan. The commission sought to reserve prime shoreline sites for what it termed "priority uses": ports, water-related industry, airports, wildlife refuges, water-related recreation, and public access to the bay (only ten miles of the entire 276-mile shore was at that time open to the public).

To carry out their recommendations and to continue surveillance of fill proposals the commissioners sought the creation of a limited regional government body. They esti-

mated the costs of acquiring 7,400 acres of submerged
land deemed of particular importance for recreation and
wildlife at between $30 million and $50 million. If it became necessary to compensate private owners of submerged land for keeping their property as open water
(rather than developing it profitably, which the commission
could prohibit), the extra cost could go as high as $285
million.

The lobbying effort against BCDC constituted the heavi-
est offensive mounted against any legislation proposed at
the 1969 session of the legislature. A conspicuous opponent
of certain provisions was Westbay Community Associates,
a firm that owns 10,000 acres of the bay and presented a
plan for development of twenty-seven miles of peninsula
shoreline below San Francisco airport, including 4,700
acres of fill. The Westbay design included residential, industrial, and recreation areas. The meetings at which Westbay tried to present its plans were "all packed with
articulate and rabid conservationists," a company spokes-
man, Warren Lindquist, recalls, and as the debate continued, the issue "became a matter of right or wrong—if you
didn't agree with the BCDC you were wrong." Another
Westbay representative, Richard Archer, told a legislative
committee plaintively that the BCDC plan was unsatisfac-
tory. "The plan should have teeth," he said, "but this one
has fangs."

Strong opposition to specific aspects of the plan came
from other firms with large holdings in or along the bay.
Against the platoons of lobbyists for the opponents, which
included the Leslie Salt Co. and the Atchison, Topeka &
Santa Fe Railway Co., the tax-exempt conservation groups
could not field even one lobbyist, since taking a political
position could cost them their tax status. Accordingly, sev-
enty conservation groups joined to form the Planning and
Conservation League, which did not claim tax exemption
and which hired a lobbyist to work for the bill.

But the main effort in support of the bill came from
thousands of concerned citizens. The Save the Bay Associa-

tion had worked uninterruptedly for three years to support
the commission, sending out communiqués on new developments, ringing hundreds of telephones, promoting the cause
on radio and television, and expanding its support from
5,000 to 22,000 dues-paying members as the crucial vote
drew near. In addition, two vigorous, deeply committed
San Mateo women, Janet Adams and Claire Dedrick,
formed the Save Our Bay Action Committee. The committee sold bumper stickers, ran a succession of full-page
ads in local papers urging citizens to "demand a halt to
122 years of destruction of San Francisco Bay," and organized charter bus trips to attend committee hearings in
Sacramento.

More than 200,000 signatures were gathered on petitions
asking Governor Reagan to support the bill. When he was
not available to receive the mass of documents, the committee, with a keen eye for publicity, strewed them—they
totaled 3.4 miles in length when volunteers had stuck
them all together—across the capitol lawn and steps, then
like true conservationists picked them all up. Meanwhile
the Sierra Club had thrown its full national weight behind
the campaign, the first urban conservation issue the club
had endorsed nationally.

Passage of the BCDC bill remained uncertain right up
to the end of the session. Volunteers continued to pack
the hearings, and letters poured in to legislators. So many
telephone calls were made in support of the bill that the
president pro tem of the senate had to plead with conservation groups to cut them off so that state business could be
transacted.

The bill passed, with its provisions for control over fill
and dredging preserved intact. There were a few modifica-
tions: most important, one reducing commission authority
over the shoreline from 1,000 feet to 100 feet back from
the water's edge. Another alteration provided that some
commission members were to be elected county supervisors
or city councilmen rather than conservationist-appointed
city and county representatives. This change reflected the

behind-the-scenes battle between "home rule" supporters and those who favored a truly regional government, responsible to the area rather than to specific communities. It was a modest but significant victory for the developers, since it gave them some election-time leverage over commission members.

The commission was given three years to select and acquire privately owned lands it believes are needed for public purposes, under the law. It has no funds for these purchases, but BCDC executive director Joseph Bodovitz hopes state or federal monies will become available. At present the annual operating budget of $200,000 is being met out of the state's general fund. But Governor Reagan does not approve of the state's footing the bill for this regional function; he believes it should be paid for by bay-area citizens. Various methods of financing are being considered, including a special bay-area income tax.

Conservationists across the country agree that this pioneering effort provides many lessons for other groups determined to protect other estuaries and natural resources, and that the plan developed for the bay furnishes an excellent pattern for management of a major natural resource. "But the battle is not now, and never will be, won," says Bodovitz. "Maintaining a high degree of continuing public interest is crucial."

XII

Our New Awareness
of the Great Web

by William Bowen

Predictions about anything much less predictable than the rising of the sun are likely to be wrong, or at least seem wrong in hindsight. So we may assume that most predictions put forward in 1937, like those of other years, would now be worth recalling only as examples of fallibility. But at least one prediction published in that year has since come to seem exceedingly perspicacious. It appeared in a book by Kenneth Burke, a literary critic. "Among the sciences," he wrote, "there is one little fellow named Ecology, and in time we shall pay him more attention."

Quite a few years passed before Burke's prophecy was borne out. As recently as 1962 the naturalist Marston Bates wrote: "Ecology may well be the most important of the sciences from the viewpoint of long-term human survival, but it is among those least understood by the general public . . ." Even a year or two ago, anyone not a biologist or a biology student could easily go for months on end without encountering any mention of ecology.

But now, almost suddenly it seems, ecology is popping out all over—the word, at least, if not the science. We meet ecology at dinner parties, in newspaper editorials, on the covers of magazines, in speeches by public officials, at gatherings of scholars in fields remote from biology, and in the names of recently born or reborn corporations (Ecological Science Corp., Ecologic Resources Corp., Ecology Inc.). At this rate, ecology can be expected to debut before long in *Playboy,* the manifestoes of student rebels, the public utterances of Edward Kennedy, subway-station graffiti, and catch-breeze book titles—*You and Ecology* and perhaps even *Ecology and the Single Girl.* There is an element of fad, of course, in this swift transformation of a mossy scientific term into a conspicuous In word. But there appears to be something much more important, too: what Kenneth Burke foresaw, awakened perception by a great many people of an urgent practical need for the kinds of information, insights, and concepts embraced in ecology.

The term ecology was coined a hundred years ago by the German biologist Ernst Haeckel. The eco-, from the Greek *oikos* (house), is the same eco- as in economics, and according to an old definition, what ecologists study is "the economy of animals and plants." In the now-standard definition, ecology is the science of the relations between organisms and their environment.

That will do as a working definition if we bear in mind that neither in nature nor in the thinking of ecologists are there two distinct compartments, organisms and environment. For any organism, other organisms constitute part of the environment. And the physical environment itself is largely created and maintained by organisms. Atmospheric oxygen, necessary to the survival of life on earth, is itself a product of life, slowly accumulated from the transpiration of aquatic organisms and terrestrial plants. A hardwood forest can maintain its stability for many centuries on end because it creates its own peculiar environment, in which seedlings of only certain plant species can grow to maturity. Recognizing that organisms and their physical environment are interacting parts of a system, an ecologist uses the term "ecosystem" to mean the community of living things and the physical environment, both together, in the segment of nature he is studying.

Ecologists study all kinds of segments, great and small.

One ecologist may investigate how various species of mites coexist in the pine-needle litter on a forest floor by occupying separate "niches," or ways of making a living. (It is a well-established principle of ecology that only one species can occupy a particular niche in any habitat.) Another ecologist may work out the intermeshed food chains of various species in a pond or a forest. Still another, a worker in the sprawling, almost unbounded field called "human ecology," may trace the paths by which radioactive substances and persistent pesticides, created by our interventions in nature, accumulate in the tissues of our bodies.

In the diverse studies of ecologists, certain basic themes keep recurring. Together, they may be regarded as the compressed wisdom of ecology.

Interdependence. "The first law of ecology," biologist Barry Commoner remarked not long ago, "is that everything is related to everything else." The continued functioning of any organism depends upon the interlinked functionings of many other organisms. Seemingly autonomous man ultimately depends upon photosynthesis for his food. The seemingly autonomous oak in the forest depends upon microscopic organisms to break down fallen leaves, releasing nutrients that can be absorbed by its roots. Interrelations between organisms are often intricate, and some obscure species provide vital linkages not at all apparent to the casual observer. The seeds of the bitterbush, an important food plant for browsing animals in arid sections of Africa, fail to germinate unless several seeds are buried close together below the surface of the soil; that happens in nature only through the intervention of a species of ground squirrel, which buries hoards of seeds and often forgets them. It is unwise ever to assume that a species is entirely dispensable.

Limitation. The saying that trees do not grow to the sky expresses another basic theme of ecology. Nothing grows indefinitely—no organism, no species. Much more commonly than non-ecologists might suppose, animal species limit their own growth: rates of reproduction respond to crowding or other signals so that total numbers remain commensurate with the resources of the ecosystem. In the over-all ecosystem of the earth, an outer limit to total animal energy is established by the amount of solar energy plants embody in organic compounds.

Complexity. When he looks closely at any ecosystem, the ecologist invariably comes upon complexity, an intricate web of interrelations. A diagram showing the movement of a single chemical element through an ecosystem can get exceedingly complicated. In the ecosystem of man, which includes institutions and artifacts that themselves impinge upon and alter the environment, the interrelations are unimaginably complex. This great web, an ecologist said, "is not only more complex than we think. It is more complex than we *can* think."

In their complexity, ecosystems exhibit some of the characteristics of complex systems that Professor Jay W. Forrester of M.I.T. pointed to in FORTUNE ("Overlooked Reasons for Our Social Troubles," December, 1969). In ecosystems as in social systems, causes and effects are often widely separated in both time and space. Accordingly, our interventions often yield unexpected consequences.

After years of spraying persistent pesticides to kill insects, we find that we have come close to wiping out a national symbol, the bald eagle: concentrated through food chains, pesticides accumulate in the tissues of eagles and certain other birds to the point of impairing reproduction. We drain Florida swamplands and learn later on that by reducing the outflow of fresh water into estuaries we have increased their salinity and thereby damaged valuable breeding environments for fish and shrimp. The Aswan Dam impounds silt that would otherwise be carried downstream, so the Nile no longer performs as richly as before its ancient function of renewing fields along its banks. The fertility of the Nile Valley is therefore declining. That is only one variety of ecological backlash from this triumph of engineering. With the flow of the river reduced, salt water is backing into the Nile delta, harming farmlands there. And in time, some authorities predict, the flow of Nile water to new farmlands through irrigation canals will bring on a calamitous spread of schistosomiasis, a liver disease produced by parasites that spend part of their life cycle in the bodies of snails.

Professor Garrett Hardin of the University of California pithily expressed the principal lesson of all this in pointing out that "we can never do merely one thing." When we intervene in a complex system so as to produce a certain desired effect, we always get in addition some other effect or effects, usually not desired. As Hardin also said: "Systems analysis points out in the clearest way the virtual irrelevance of good intentions in determining the consequences of altering a system."

Ecologists are accustomed to looking at nature as a system, and if we had paid more attention to them we might have been spared a lot of backlash. In trying to reduce insect damage to crops, for example, we might have made more use of specific biological or biochemical means of control and less use of persistent broad-spectrum insecticides. We might now, accordingly, have more birds in our countryside and less DDT in our streams—and in some places, fewer harmful insects in our fields.

The recurrent themes of ecology run counter to some old ways of perceiving and thinking that are deeply ingrained in the prevalent world view of Western man. We believe in limitless growth (or did until recently); ecology tells us all growth is limited. We speak (or spoke until recently) of man's "conquest" of nature; ecology tells us we are dependent for our well-being and even survival upon systems in which nature obeys not our rules but its own. Our scientists and engineers, and our social scientists too, proceed by isolating and simplifying; ecology tells us to heed existent complexity and patiently try to trace out its strands. In a sense, then, ecology is subversive. The ecologist Paul B. Sears, a few years ago, called it "a subversive subject," and the editors of a recent compilation of essays on the ecology of man entitled their book *The Subversive Science*.*

In the recent popularity of the word "ecology," therefore, we may be witnessing a sign of momentous historical change. Alterations in the ways men perceive and think about reality lead to alterations in the goals and modes of action. It is too early to tell whether an enduring shift to ecological ways of perceiving and thinking is now in progress, but if it is, the effects will surely be beneficial, on balance. Ecology can help us cope with the environmental ills that beset us, if only by enabling us to avoid bringing on new unintended consequences in trying to remedy old unintended consequences.

Less obviously, ecological thinking can help us cope with the social ills that also insistently press upon us. The social sciences are proving to be inadequate as guides to policy, and the inadequacy is inherent in the prevalent methods and mind-sets of social scientists. In general, they have aped the successful methodology of the physical sciences, but in the study of complex social systems, simplification too readily slides into oversimplification. The social sciences would benefit greatly—and so would we all—by borrowing from the ecologists their willingness to accept and try to puzzle out complexity and their habit of sustained, open-eyed observation of what actually goes on.

All this may be too much to expect. But it seems possible, to take a cheerful view, that in 1980 or 2000 Americans will be better off in their physical environment *and* their social arrangements because, at the beginning of the 1970's, Kenneth Burke's prediction came true.

---

* *Edited by Paul Shepard and Daniel McKinley and published by Houghton Mifflin. The book is a rich trove, hard digging in places, but worth it.*

XIII
How to Think About the Environment

by Max Ways

Slowly as a poisoned snail, public concern over the physical environment crept forward during the first half of this century. In the Fifties anxiety began to quicken. In the Sixties the environment became one of the major topics of American discussion. In the last few months it has been almost impossible to look at a newspaper or a TV set or to talk to a stranger on a plane without encountering some example of dismay over what we are doing to the world around us.

Yet snails, along with all the other species—notably including man—that live on or near this once fair land, continue to be poisoned or choked or pushed around or otherwise imperiled. If, as seems quite likely, public anxiety mounts to ten times its present level, we will not necessarily succeed in correcting our present lethal course. Alarm about the environment, fully justified by the facts and long overdue, is a required precondition of reform. But alarm, by itself, puts out no fires.

It is not hard to imagine how disgust with our mishandling of the environment could become, as in some quarters it already has, just one more taunt against "the sick society," one more whip in the orgy of self-flagellation that has seduced so many professional observers of this pseudo-penitential time. Unless guilt over what we have been doing is accompanied by a stronger sense of what to do instead, U.S. morale may sink to the point where we will be unable to cope properly with the environment—or anything else.

This book is devoted to the grave challenge that arises from American abuse of the natural resources of air, water, and land, from our failure to honor the terms of partnership in which men coexist with other forms of life, and from our neglect of order and design in our artifacts, especially our cities. The primary intent is not breast-beating or doom-saying. Americans are generally already aware that progress has, somehow, been ambushed by its misbegotten children, the unintended by-blows of the modern lust to know and to do.

American business, since it organizes and channels a high proportion of the total action of this society, has been and still is deeply implicated in depredations against the environment. Any way out of the present mess will have to rely heavily upon business for resources, for innovation, and for leadership. The political and social ground rules within which business is conducted will have to be drastically amended. Important corrections will have to be made in the terms of economic calculations that now skew business decisions in destructive directions. In short, business, along with all other major functions of society, will have to change if Americans are to achieve a better environment.

Other chapters in this book deal with specific aspects of environmental problems. But a successful reform of our patterns of action will require much more than piecemeal dosage of technical antidotes for one environmental poison after another. This chapter essays to view the challenge in the round, to probe for the root common cause of such disparate symptoms as traffic jams and vanishing lakes, to establish a perspective between material blight and material progress, to sketch the character of new moral, intellectual, political, legal, social, and economic patterns that must soon be evolved. And beyond this tall order lies the inescapable question of how an effective social program of cherishing the environment might fit with the future of the long march toward individual freedom, and whatever else we mean when we speak of "the quality of life."

Winston Churchill in October, 1943, made in the House of Commons a gemlike speech on an unlikely topic for a time of total war: the relation between architectural design and human behavior. The old House had been blasted by a German bomb, and the question was how it should be reconstructed. A more spacious chamber, perhaps? One with a semicircular arrangement of comfortable chairs and useful desks, such as many of the world's parliaments enjoyed? Churchill thought not. What he wanted, essentially, was a replica of the old House with its rows of facing benches that symbolically expressed the party structure, emphasizing the contrasting roles of the Government and the Opposition. The new House, like the old, should have far fewer seats than there were members, so that in an ordinarily ill-attended session speakers would not be discouraged by addressing empty benches. On great occasions, members flocking in would be crowded standing in the aisles, thus encasing debate and decision with visible signs of gravity and urgency. No desks, for members might bang them as had the rowdy French whose parliamentary institutions, he did not need to recall, had proved less than stable. Churchill summed up these minutiae with a thought that is far from trivial, "We shape our buildings, and afterwards our buildings shape us."

Rare in our civilization are such instances where a public man (or, for that matter, a private man) evinces a thoughtful, constructive concern for the indirect human implications of material design. Yet how men deal with things—whether a room, a city, a river, or the great biosphere itself—both discloses and thereafter influences the way they see themselves and the way they deal with one another. Consciously or unconsciously, we make day-by-day decisions that affect the material world. These decisions have an immediate moral and aesthetic content that should be taken seriously, even if a deteriorating environment were not a long-range threat to prosperity, health, and the continuation of the race.

Survival is not the only—and perhaps not the most impelling—motive for environmental reform. People have a way of brushing off warnings that their farmlands, their lungs, and their life expectancies are being slowly eroded. The threatened penalties, though severe, seem to lie in the remote future. But inhabitants of drab and chaotic cities surrounded by a despoiled countryside are punishing themselves here and now not only by material disadvantages, but by the absence of a precious immaterial quality. Life is presently diminished, it loses point and relish and sense of direction when it is spent amidst a haphazard squalor that God never made, nature never evolved, and man never intended.

A youth smashes a window or sets fire to a house or mugs an old woman on a street. How much have we nurtured such disorders by the disorders of the American scene? It doesn't *look* as if anybody around here cared.

Although environmental issues do have a grave moral content, there's little sense in the tendency to present the case in the dominant art form of a TV horse opera. This isn't, really, a confrontation between "the polluters" and the good guys in the white hats. Nevertheless, casting for the villainous roles proceeds briskly. "Greed" is to blame. "Man, the dirtiest animal," is to blame, especially because his numbers are increasing. "Technology" is to blame—and this charge, as we shall see, contains much truth, though far less than the whole truth. "Capitalism" is to blame. "The poor," who throw garbage in the streets, are to blame. "Democracy," which seems unable to find remedies, is to blame. And, of course, "the establishment," everyone's goat of atonement, is to blame.

In general, the nomination of villains follows the familiar pattern of dumping the ashes of contrition on somebody else's head. A Columbia law-school senior this year was reported to have boasted that he told recruiters for law firms he "would not defend a client who was a polluter —and most of the clients who pollute are the big ones," a remark indicating that even law-school seniors may have something to learn.

For all men are polluters—and all living Americans are big polluters. The greedy and the ungreedy alike befoul the air with automobile exhaust fumes, the humble 1960 jalopy contributing somewhat more poison than the arrogant 1970 Cadillac. So long as our laws and habits of land use foster chaos, the homes of saints will aggress as rudely upon nature as the haunts of sinners. Who killed the rivers of Illinois by extinguishing perhaps forever their ability to cleanse and renew themselves? The effluents of big industries did a substantial part of the damage. Sewage from towns did part. But most of the damage to the rivers of Illinois came from farms onto which decent and well-meaning "little" men, in the pursuit of the legitimate aim of increasing crop yields, poured nitrogen fertilizers. The result bears the mellifluous name of "eutrophication": algae, slimy green gunk, rampantly feed upon the fertilizer drained into the rivers; the decay of dead algae consumes so much of the available oxygen as to destroy the bacterial action that once cleansed the rivers of organic wastes.

At the root of our environmental troubles we will not find a cause so simple as the greed of a few men. The wastes that besmirch our land are produced in the course of fulfilling widespread human wants that are in the main reasonable and defensible. Nor will we find capitalism at the root of the trouble. The Soviet Union, organized around central planning, has constructed some of the most terrifyingly hideous cityscapes on earth, while raping the countryside with strip mines, industrial pollutants, and all the other atrocities that in the U.S. are ascribed to selfish proprietory interests. Aware, as well they might be, of American environmental mistakes in handling the mass use of the automobile, Russians keep saying they will do it better; but today, as automobiles become more numerous in the U.S.S.R., it is hard to find in city or highway planning, in automobile design, or in any other tangible area signs that they are in fact better prepared for the automobile onslaught than the U.S. was in 1920.

The Japanese, though their basic culture lays great stress on harmony between man and nature, are not handling their environmental relations significantly better than the Americans or the Russians. Japan's economy, combining private enterprise with government central planning, seems able to do anything—except cherish the material beauty and order that the people value so highly.

If we wish to think seriously about the environment, we have to give up indulgence in barefoot moralism and the devil theory of what's wrong. We have to identify a root cause that explains the environmental failures of systems as different as the American, the Russian, and the Japanese. Obviously, all three are high-powered industrialized, technologized societies, and our quest for a root cause can start by tentatively picking technology as the villain.

Despite billions of words on the subject, we still underestimate the magnitude of technological advance and its implications. Thirty years ago R. Buckminster Fuller found an apt way of expressing what had occurred. He calculated the total energy generated in the U.S. as equal to the muscular energy that would be generated if every American had 153 slaves working for him. Today a similar calculation would indicate about 500 "slaves" for every American man, woman, and child.

These slaves enable us to increase our own mobility hundreds of times and to toss around incredible masses of materials, altering not only their location and external shapes but their very molecules. Excluding construction, earth moving, and many other operations, the U.S. economy, according to one estimate, uses 2,500,000,000 tons of material a year. That's nearly thirteen tons per person.

These figures explain a lot of environmental woes that are otherwise mysterious. Although our cities are not more densely populated, they produce more maddening and wasteful congestion than any cities of the past. Our crowding is not basically a matter of too many human beings to the square mile but of the enormous retinue of energy and material that accompanies each of us. Like King Lear with his hundred riotous knights and squires, we strain the hospitality of our dwelling space, and from our situation, as from Lear's, much grief may follow.

Two hundred million of us are bustling about the U.S., every one sheathed in a mass-and-energy nimbus very much bigger, noisier, dirtier, smellier, clumsier, and deadlier than he is. The paper, plastics, scrap, ash, soot, dust, sludge, slag, fumes, and weird compounds thrown off by the mass-and-energy nimbus exceed by many magnitudes our own bodily wastes. If ten billion mere people, sans technological nimbus, inhabited the U.S., they could not create more congestion, blight, and confusion. The three million high-technology U.S. farmers put more adverse pressure on their land and rivers than the hundred and fifty million low-productivity peasant families of China put upon their land and rivers.

How should a city be designed and its circulatory system arranged to accommodate a people that employs energy and mass at present American levels? The past offers only wisps of inspiration, but no usable models. Consistently we have failed to face the sheer physical challenge of the contemporary city, assuming that old urban forms would be adequate if we amended them a little to meet one crisis after another.

Along with all kinds of congestion, our cities produce a paradoxical effect of isolation and desolation. Not rationally shaped for the needs of this society, the cities may be shaping us toward irrationality. Frequently mentioned in environmental circles these days is a research project carried out by John Calhoun at Bethesda, Maryland. He placed Norway rats in a closed area ample for the original population. As they multiplied, the crowded animals, though well fed, developed most distressing psychoses, which, out of a decent respect for the privacy of rats, will not be here detailed. Many who have heard of this project see a close parallel with our cities.

But the analogy is not quite true to the situation of contemporary urban man. It would be better to find a strain of rats each one of which had the services of a half-house-broken elephant to do its work, run its errands, and cater to its wants. In an ill-organized space these lordly rats, even if they did not multiply, might go crazier quicker than did their cousins in Bethesda.

People who center their anxiety on "the population explosion" see the challenge much too narrowly. In the U.S. and other advanced countries, population has been increasing less rapidly by far than the explosive acceleration of the total energy and total mass deployed. If the population declined and technology continued to breed, without any improvement in the arrangements for its prudent use, a small fraction of the present U.S. population could complete the destruction of the physical environment while jostling one another for room.

We come, then, to the question of whether a headlong retreat from technology would be the right strategy. This option needs to be honestly appraised, not toyed with as it is every day by nostalgic romanticists wiggling their toes in secondhand memories of Thoreau's Walden Pond.

The casualties of a withdrawal from technology would be heavier than many suppose. Everybody, of course, has his own examples of unnecessary technologies, unnecessary products, unnecessary activities. But because we are, thank God, diverse in our wants, the lists do not agree. The man who has since childhood said to hell with spinach has a ready-made response to the news that a high incidence of "blue babies" recorded in Germany has been attributed to heavy use of nitrogen fertilizer on the spinach crop. But other consumers will have good reasons for wishing spinach yields to increase. We will not improve our environmental situation by recommending a technological retreat on the basis of what each of us considers the superfluous items in the households of his neighbors.

To be effective in protecting the environment a technological retreat would extend over a wide front and go back a long, long way. A century ago we had already slaughtered the bison, felled the eastern forests, and degraded the colonial cities. Retreat to the 1870 level of technology, while not giving long-range protection to the environment, would place the median American standard of living far below the 1970 poverty line. Among the consequences of such a retreat would be the closing of 75 percent of the present colleges and most of the high schools. We would give up not only automobiles and airplanes but also mass education and social services. Grandpa would return to living in the abandoned hencoop.

Since we are not going to choose such a retreat from technology as a deliberate social policy, sheer practicality forces us to seek another way out. In that quest we have to ask seriously why the U.S. and all the other advanced countries have failed so dismally in handling the unwanted effects of technology.

Modern technology did not spring out of the void. It did not well up simultaneously in all the world's peoples. It appears first in European culture, and, although it is now disseminated over the whole globe, its main generative fonts remain to this day Western.

The Western origin and leadership in technology, the main agent of environmental destruction, inevitably raises uncomfortable questions about Western culture itself. The Judeo-Christian religious formation is not essentially "anti-nature," as some angry men now aver. But in contrast to Oriental religions, it does sharply separate its idea of God from its idea of nature and does look upon man as having a special relationship with the Creator and a unique place within creation.

The Western tendency to objectify nature—to see it "from outside"—is undoubtedly responsible for much arrogant and insensitive handling of the material world. But it ought not be forgotten that this same attitudinal "separation" of man from nature forms the basis of man's ever increasing knowledge of nature. In recent centuries, especially, Western man has empirically confirmed his ancient notion of himself as unique among the creatures: no other species possesses a glimmer of his ability to learn about nature and to operate, for better or worse, upon it.

Surely it can be no accident that four centuries of science are attributable almost entirely to Western culture. Extending the pattern of Western religion and philosophy, which had drawn sharp distinctions between ideas of God, of man, and of nature, the scientific method began to separate one aspect of nature from another for purposes of study. This superlatively effective way of discovering solidly verifiable truths tends, precisely because it is sharply focused, to ignore whatever lies outside its periphery of attention.

Science, seeking only to know, is guiltless of direct aggression against the environment. But technology, devoted to action, feeds ravenously upon the discoveries of science. Although its categories are not the same as those of science, technology in its own way is also highly specialized, directed toward narrowly defined aims. As its power rises, technology's "side effects," the consequences lying outside its tunneled field of purpose, proliferate with disastrous consequences to the environment—among other unintended victims.

Modern Western man has advanced the principle of separation or differentiation also in areas of life, such as psychology and politics, that are seemingly remote from science and technology. These trends, too, have contributed to our environmental diseases.

The undifferentiated human mass, say of ancient Egypt, has been replaced by modern men who regard themselves —and in fact are—highly individuated. The long trend to individualism, which has Greek as well as Judeo-Christian origins, has sharply accelerated during the modern centuries. One of its aspects, democracy, is based on the assumption that the diverse wants, skills, interests, and opinions of individuals should not be ignored or rudely aggregated from above, but must be somehow coordinated from below. The latter process is clearly the more difficult, especially when applied to such large questions as how to protect the physical environment from human misuse.

The pharaonic society employed its most potent technology, irrigation, on the premise that everybody shared a common desire to eat. The knowledge required to operate the system was closely held in a group of priestly intellectuals, and the decision-making power was concentrated in the will of Pharaoh. With its technology under unified control, with few conflicts or complications arising from a diversity of skills, interests, rights, and powers within the community, the pharaonic society could maintain for centuries a stable, harmonious relation with nature and could also achieve stylistic and functional coherence within its man-made environment—such as it was.

One can hear today in environmentalist circles half-serious remarks that every city needs a king and every country an all-powerful planner to unify decisions affecting the environment. Such suggestions underestimate the human cost of the reversal, as do proposals for a retreat from technology. We will not voluntarily abandon the view that society should be made up of highly individuated men pursuing their own aims by their own lights.

We have permitted the free combination of individuals on the basis of shared specific aims. By means of such groups, mainly corporations, we have organized and stimulated technological advance, matching techniques to particular group aims. Though this pattern all too often ignores the undesirable side effects of its single-minded thrusts, it fits so closely with the evolution toward human diversity and freedom that we would shrink from a return to the pharaonic kind of harmony and stability. We are not fellahin, and the road back to that condition might be more arduous and more disorderly than the road we have traveled.

Western culture has never denied that a society stressing the individuality of its members needs the restraint and to some degree the positive leadership of government. But the character of government has also been affected by the trend toward differentiation. The Lord's anointed, with unspecified and even "absolute" power, has been split up into sharply segregated bureaucratic functions.

These, too, generate undesirable side effects. A highway department's mission is defined by statute and by specific appropriations. As it goes about its assigned task of building the most road for the least measured cost, it rips up neighborhoods and landscapes, creating enormous social disutilities that never get into the department's benefit-cost calculations. A sanitation department, told to dispose of garbage, may tow it offshore and dump it. When the refuse washes back upon the beaches and into the estuaries, the problem belongs to some other department. Or the specialists in solid-waste disposal may burn trash and garbage in places and in ways that transfer the pollution to the air.

Fragmentation of modern government occurs even in "totalitarian" countries. Administration of the Soviet economy is divided among fifty-odd ministries for the sake of efficiency. If a paper mill is needed, the men told off for that responsibility look around, like any capitalist, for plentiful timber, plentiful water, and cheap electric power. One paper mill was placed on the shore of beautiful Lake Baikal because the protection of this unique body of water lay outside the field of assigned vision of the men in charge of paper production. They were not being "greedy" or even "stupid" in the ordinary meaning of those words. They were wearing the blinkers of concentration, using the great Western device of fixing attention on the job at hand, of dealing intelligently with one segment of reality at a time.

Though the principle of segregated attention proves

gloriously successful—in research, in work, and in government—it can collide disastrously with the principle of unity. For each man is a unit though his skills and wants may be various. A society is a unit as well as a multitude. Nature, most marvelously connected throughout all its diversities, is a unit. Violation of these unities invites penalties and poses formidable tasks of reintegration.

Here we come to the root cause of our abuse of the environment: *in modern society the principle of fragmentation, outrunning the principle of unity, is producing a higher and higher degree of disorder and disutility.*

How can balance be restored? Since it is profoundly unrealistic to believe that we will or should retreat from such bastions of diversity as science, technology, and human individuality, then we have to seek improved methods of coordinating our fragmented thought and action.

During recent centuries, institutions of coordination, though lagging behind diversity, have not stood still. In economic affairs the market performs, albeit imperfectly, a stupendous job of mediating disparate wants, skills, resources. Government, amidst its bureaucratic fragments, has not completely lost the notion that it is supposed to serve such unitary purposes as "the general welfare." Specialized knowledge has a medium of transfer in the great modern webs of information, particularly the universities where all the sciences meet even if they do not fluently communicate.

How might such integrative agencies as market, government, and university be used, separately or in combination, so as to minimize the damage that fragmented action now does to the environment? This is the question on which the chance of actual reform, as distinguished from alarm and breast-beating, depends.

In two areas, air and water pollution, a moment's reflection should convince anybody that the market, as now set up, is rigged against the environment. A hundred and fifty years ago it was almost unimaginable that clean water, much less clean air, could become scarce in the

U.S. economy. Rightly, these resources were then considered common property and used without charge. The price of everything else the economy uses—land, minerals, food, labor, time—becomes dearer. But clean air and water, though now precious, are still left out of the pricing system, still free of charge.

Because the market has failed to keep pace with changing economic reality, the pricing system, expressing relative demand and supply, works against the conservation of clean air and water. A manufacturer is under great pressure to offset rising labor and material costs by developing new techniques. He has been under no comparable pressure with respect to clean air and water. Not surprisingly, techniques for conserving these resources have developed very slowly. The effect of omitting free resources from the pricing system is to make the economy as a whole pay a huge subsidy to those activities that put above average pressure on free resources. In short, we are now providing a huge, unintentional market incentive to pollution.

The most direct and logical way of getting clean air and water into the market system is by a federal tax graduated in respect to the quantity and undesirability of the pollutants. Such a tax, escalating over the first five or ten years so as not to destroy industries whose cost structures are based on the present system, would stimulate the development of antipollution techniques.

Taxes on the abuse of water and air would not replace the present trend toward stricter antipollution measures enforced by police power. Radioactive wastes, for instance, can be dangerous in very small quantities because they concentrate as they move up the food chain. The strictest control of such wastes is required—and may prove expensive. Nuclear power will be better able to absorb such costs if its competitor, fossil fuel, is forced to pay for the clean air and water it displaces. By such combination of government police power and taxing power we can turn the market toward protection of the environment—or at least achieve its "neutrality."

Correcting the market is much more difficult in that growing class of cases where the bad environmental side effects do not occur until the product is in the hands of the consumer or even until after he has disposed of it. It is by no means clear that automobiles, for instance, now carry taxes equivalent to the true social costs incurred by their use and disposal. If we become serious about the preservation or restoration of public transport in American cities the first step would be to make sure that public policy is not subsidizing the automobile.

Still more difficult to deal with is the product that is innocent until it interferes with some technique of protecting the environment. Many plastics give trouble in this indirect way. The polyvinyl chloride bottle causes no problems unless it is burned in a trash incinerator that is equipped with a scrubber designed to catch soot and fly ash. The burning PVC causes hydrochloric acid to form in the scrubber, destroying its metal casing. Some companies that hoped to sell more scrubbers for smaller incinerators have given up because they cannot guarantee their devices against the increasing incidence of PVC in trash. A small tax based on the nuisance side effect of certain plastics would either drive them off the market or encourage a new technology that abated the nuisance. As technology advances into more and more esoteric compounds, each carefully designed for a particular use, protection of the environment will require public policies that force innovators to pay more attention to the side effects of their products.

With gratifying frequency and emphasis, business spokesmen these days are expressing their determination to exercise greater care of the environment. If business greed lay at the root of our environmental troubles then this repentance would itself signal the great turnaround. In fact, a more sensitive and socially responsible business attitude will be of very limited help—unless it is accompanied by new ground rules. Under the present set of rules, if one corporation is environmentally a good citizen, incurring heavy costs to fight pollution, and if its competitor operates on an environment-be-damned basis, then the first corporation will be punished and the second rewarded. The market will practice selection against the environment.

Instead of getting on with the formidable job of rewriting the rules, public discussion wastes time and energy on irrelevant questions, such as how much of business profit should be diverted to environmental betterment. The problems have become so huge that we would not necessarily make a dent in them with *all* the profits of American business.

Henry Ford II expressed this point cogently. He first noted that, "Hardly anyone disputes the proposition that service to society requires at least a short-run sacrifice of business profit." Then Ford disputed that consensual proposition. "This point of view," he said, "may have been tenable in the past. As long as public expectations with respect to the social responsibilities of business were relatively narrow and modest, business could pass muster by sacrificing only a little of its short-run earnings. Now that public expectations are exploding in all directions we can no longer regard profit and service to society as separate and competing goals, even in the short run. The company that sacrifices more and more short-run profit to keep up with constantly rising public expectations will soon find itself with no long run to worry about. On the other hand, the company that seeks to conserve its profit by minimizing its response to changing expectations will soon find itself in conflict with all the publics on which its profits depend."

Ford recommended not a middle course but an entirely different approach. Instead of regarding profits as competing with public expectations, such as environmental demands, he argued that business should look upon the rising public standards as opportunities for profit. "We have to ask ourselves what do people want that they didn't want before, and how can we get a competitive edge by giving them more of what they really want?" Ford showed he was fully aware that in many cases governmental action would be required to translate the public's heightened desire for a better environment into effective market demand. New antipollution standards set by law, he said, had stimulated efforts by the automobile companies to reduce noxious emissions from cars.

Ford's and all other realistic approaches to environmental problems imply future government action very different in kind and quality from the typical bureaucratic methods of the past. Some businessmen's continued opposition to government action in protecting the environment can be explained—if not defended—on the ground that their past experience with government controls does not inspire confidence. Businessmen rarely contact "government" in any general sense. They confront one bureaucrat after another. Because the bureaucracy is even more sharply fragmented than business, there is often a good reason for a businessman to conclude that the typical bureaucrat's view of the general public interest is narrower than his own.

The beginning of a shift from fragmented to integrative government action can be illustrated by a brief look at the history of a field where concern for the physical environment and concern for the social environment overlap. Generations of indignation about "the slums" came to a head in the Thirties. Crime, ignorance, disease, unemployment and, of course, poor housing were all among the reasons for anxiety. Governmental antislum action concentrated on one element, housing. Here was a material object that we surely knew how to produce in quantity. Without much study it was assumed that new public housing would somehow lever upward all the other adverse conditions of the slums.

So a large, competent, and militant bureaucracy was assembled on the narrow front of public housing. Over

the years it went from triumph to triumph, if you measured by the size of the appropriation and the number of
units built. People began to notice, however, that this
program was not delivering, in human terms, what it had
promised. The poor displaced by slum clearance were in
many cases worse off than before. The new projects, more
grim and sterile than the old slums, did not produce falling
crime rates, better health, and improved school performance. Typically, the housing bureaucrats and their supporters brushed off these observations as inspired by reac-
tionary politics and, inevitably, private greed.

A day came when the inadequacies of the narrow approach could no longer be ignored. More general terms,
like "urban renewal," began to replace "slum clearance"
and "public housing." In 1966 evolution of the governmental approach took a big step forward in the Model
Cities program, which envisions a coordinated attack on
such disparate fronts as medical care, educational enrichment, improved police work—and housing. Moreover, the
Model Cities program calls for a decision-making structure
in which people who live in the neighborhoods involved
will have an important voice in determining what is done.

Because it runs against the familiar grain of fragmentation and bureaucratic control by the specialized experts,
the Model Cities program breeds problems and conflicts.
More accurately, it makes explicit and open those problems
and conflicts that the old public-housing approach, eyes
fixed straight ahead, would have ignored. "The community," now assumed to have its own values about its living
arrangements, has been given an integrative role; it is
forced to think in terms of priorities and of the relationship between, say, health and education. As a decision-
maker, the community sometimes asks questions that
would not have occurred to experts working within dis-
crete programs. The broad approach puts pressure on the
social and environmental sciences to coordinate their
specialized knowledge and to undertake investigations of
what needs, in a practical sense, to be discovered.

There will be many cases where a high measure of integrative action can be achieved with government playing only a secondary, though necessary, role. One of the most fascinating innovations in protection of the environment is on trial just northwest of Baltimore, in the area formed by the Green Spring and Worthington valleys.

These 45,000 acres are inhabited by well-to-do and rich landowners who breed horses and cattle in one of the loveliest landscapes to be found in the U.S. The appearance of the valleys had changed little in this century, but the handwriting was on the hillsides. New highways had made the area more accessible to Baltimore. New real-estate developments pressed in from three sides. There was a wide and deep consensus among the area's 5,000 families that they would like to keep it the way it was. Despite their wealth, they doubted that they would be able to do this. Some landowner would sell to a developer. Reading this as the beginning of the end, another and another would sell. They would do so not out of greed, but because the system as it actually functioned seemed to give them no other real choice. Soon the valley floor would be covered. People who had paid premium prices to live "in the beautiful Green Spring Valley" would discover that they were living in a replica of ten thousand other undistinguished suburbs.

The valley landowners, banding together, sought advice from Dr. David A. Wallace, who had planned Charles Center in midtown Baltimore. To make an ecological study he called in Ian L. McHarg, a landscape architect out of the University of Pennsylvania. Together they worked out a "Plan for the Valleys," described as a chapter in McHarg's book, *Design with Nature*. The plan's most striking characteristic is its contrast to the purely negative approach of most conservationists. It accepted change for the area not only as inevitable but as socially desirable. Wallace and McHarg, estimating that unplanned development would generate $33,500,000 in profits, sought to devise a way of orderly development that would add as much.

The unique aesthetic asset of the area was the pastoral scene on the valley floors. There, housing subdivisions would be prohibited. Development could occur in rather dense clusters on some of the area's plateaus, now sparsely inhabited. Within thirty years, population of the area might rise from 17,000 to 110,000 or more. By 1980 more than $40 million would be added to total value.

The big jump, of course, would be in the price of development land on the plateaus. The plan calls for some of this profit to be piped down to the valley-floor owners whose abstention from sales to developers would cause the price rise on the plateaus.

Whether this plan will work remains to be seen. But it is the kind of effort that should provoke study all over the U.S. The essential elements of the plan are (1) the wishes of the private owners; (2) their recognition that the housing needs of other people require great change in the area; (3) the combination of ecological and socio-economic planning principles; and (4) support of the plan by government.

"Plan for the Valleys" runs counter to many ingrained ideas of "absolute" private-property rights and untrammeled market action. The main purpose of institutions based on these ideas is to widen freedom of choice. As our technological power rises it becomes more and more obvious that the use and value of one man's land depend on what his neighbor does with his land. If this new reality is not reflected in our laws and other social institutions then actual freedom of choice diminishes. The wholesome "fragmentation" represented by private landownership will defeat itself in the absence of balancing institutions that can coordinate social action in respect, say, to the preservation of a beautiful environment, a unity that gives value to the ownership of its parts.

Multiplication of laws setting forth what can't be done will be ineffective unless we invent legal, economic, and social devices through which we can decide what *should* be done. In the absence of such decision structures few people will ever develop Winston Churchill's sense of the connection between material design and human purpose.

Compare the constructive principles of the "Plan for the Valleys" with the negative and accusatory tone of much recent discussion. At a meeting in New York recently, various environmental ills, including the appalling mess of the Jersey Meadows, were excoriated. Said one of the guests, a businessman: "Jersey Standard's officers should have been shot for putting a refinery there in the first place." Another guest asked: "Where should they have put it—in the Rocky Mountains?" The businessman was appalled by this sacrilegious suggestion, but he refused to deal seriously with the question of a refinery's location.

So do most conservationists. Everybody wants ample electric power, for instance, but more and more communities are prepared to resist the presence of a power plant. Decisions on plant location are being made in a basically disorderly way with each fragmented community interest in turn poised against the fragmented interest of the power company. More and more of these cases are getting into courts. But the courts, operating in the narrow dualism of adversary proceedings, are hardly in a position to say where a power plant ought to go. In this decision system the most apathetic or careless community would get stuck with the power plant or refinery—and this might be exactly the worst place to put it from either a business or an environmental viewpoint.

Obviously, a high-technology society needs, and its government should provide, forums for the rational resolution of such questions. Carl E. Bagge, a member of the Federal Power Commission, argues that regional planning bodies should become the forums for deciding on such questions as the location of new power plants and power transmission facilities.

Better handling of the environment is going to require lots of legal innovation to shape the integrative forums and regulatory bodies where our new-found environmental concerns may be given concrete reality. These new legal

devices will extend all the way from treaties forbidding
oil pollution on the high seas down to the minute concerns
of local government. But the present wave of conservationist interest among lawyers and law students does not
seem to be headed along that constructive path. Rather, it
appears intent on multiplying two-party conflicts between
"polluters" and victims.

When we read of some environmental atrocity—a sonic
boom, a baby bitten in a rat-infested slum, a disease caused
by polluted air—our sympathies instantly go out to the
victims, just as our sympathies go out to those hurt in
automobile accidents. This example should give us pause.
The damage suit as a legal remedy in automobile accidents
has clogged the courts and imposed on the public a $7-
billion annual bill for liability insurance premiums. This
huge cost contributes almost nothing to highway safety.
For a fraction of the dollars and the legal brains drained
off by damage suits we could have produced better highway codes and better regulations for car safety—and also
provided compensation for the victims of a diminished
number of accidents. If environmental law follows the
dismal pattern of automobile tort cases, every business and
perhaps every individual will be carrying insurance against
pollution-damage suits. An army of pollution chasers, hot
for those contingent fees, will join the present army of
ambulance chasers. None of that is going to do the environment any good.

From the civilizational standpoint, the expansion of the
law of torts was a magnificent advance over the blood-
feud, the code duello, and the retaliatory horsewhip. But
out of respect for this achievement of our ancestors we are
not required to go on multiplying damage suits ad infini-
tum, while ignoring the need for new legal forms more
relevant to the problems of our own time. This is not
intended to suggest that environmental tort cases should
have no place in future law. It is meant to express the
hope that such suits will be exceptional and that the main
line of legal development in respect to the environment

will break (if conservationists can forgive the metaphor) new ground.

The web of communication could do a much better job than it is doing in helping to coordinate society's fragmented actions. Journalism, enslaved to its own outworn traditions, is eager to report those environmental concerns that are reduced to simple and dramatic "confrontations," but journalism is insensitive to the complexities of city planning or of market reform to reduce pollution.

A key contribution to the environmental future can be made by the university, the most significant institution in the whole communication network. Indignation concerning the environment is now at a very high pitch among students and faculty. Not all of this emotion, however, is translated into efforts within the university to balance specialized fragmentation with integrated studies.

An interesting view of faculty attitudes toward the environment was elicited from Robert Wood, Undersecretary (and briefly Secretary) of Housing and Urban Development in the Johnson Administration and now director of the Joint Center for Urban Studies of M.I.T. and Harvard. On his return to academia, Wood found that, throughout the faculties, interest in environmental affairs had suddenly become emotionally intense. "They are a little like the atomic scientists after Hiroshima," he says. "They had assumed that science was automatically improving the world. Confronted with contrary evidence, they feel guilt. But they tend to become impatient when their condemnations of what is happening are not immediately followed by correction."

Many faculty members who are most indignant about the environment would be unwilling to direct their own research or teaching effort to environmental questions. Urban-affairs centers and institutes of ecology are proliferating on campuses but in many cases they are not allowed, because of the jealousy of the entrenched disciplines, to give credit courses or degrees. In the academic structure, such interdisciplinary institutes are looked down upon much as a mongrel would be regarded at a show sponsored by the American Kennel Club.

U.S. society is going to need tens of thousands of "integrators," men who can handle environmental material from several natural sciences in combination with material from several of the social sciences. These men will utilize very high technologies, such as computers and space satellites, to diagnose and cure the side effects of other technologies. Tomorrow's integrators, moreover, must be able to deal with broad questions of human value, purpose, and law that lie beyond (and between) the sciences. The universities that produced the specialists who taught us how to take the world apart will now have to train the men who will take the lead in putting it together again.

Environmental damage is so widespread and is continuing so rapidly that there is a serious question as to whether we can afford reform—a question that is not necessarily answered by the glib truth that we cannot afford to go on as we are. Chapter IV deals directly with economic aspects of environmental problems. Here, however, it is appropriate to place efforts to improve the environment within the perspective of this society's economic future.

If certain twentieth-century trends such as population growth and, especially, the enlargement of the per capita mass-and-energy nimbus are simply extrapolated into the future, it's obvious that at some point we will destroy ourselves by consuming the earth. But these rates may not soar on forever. After ten years of falling U.S. birthrates, it has become possible to believe that the U.S. population may stabilize between the years 2000 and 2020 at not much above its present level, as a few demographers have predicted.

Limits of growth are also in sight for the more important rate of mass-and-energy used. The heavy environmental pressures come from agriculture and manufacturing (including mining). We are already producing more food than we consume and more than we would need to feed all the hungry in the U.S. The total value of manufactured goods will probably continue to rise for several decades, although substantial reductions in this demand could result from better environmental policies. Many things (e.g., the second family car, the second home) that we now buy are made "necessary" by wasteful environmental arrangements. The U.S. will probably reach saturation in manufactured goods in any event at some point in the next fifty years, if only because the time to use all the things we buy is becoming scarce.

Meanwhile, this economy will be very hard pressed to keep up with its increasing needs on the "services" front. A society that is both highly specialized and rapidly changing requires, as ours has already demonstrated, an elaborate "nerve system," employing millions, to maintain its cohesion and determine its direction. Among the elements of the "nerve system" are education, communications, law, finance, etc., which burn little fuel and consume small tonnages of materials.

The probability that gross pressure on the environment is due to stabilize does not of itself constitute a ground for optimism. It merely indicates that our prospect is not hopeless, and that by a huge and intelligent effort we might reverse the present devastation.

Whether that effort will be made depends primarily on how we think about the challenge. We did not get into this mess through such vices as gluttony, but rather through our virtues, our unbalanced and uncoordinated strengths. If we do not succeed in bringing under control our new-found powers, the failure will be attributable to the father of all vices, inattention to the consequences of our actions.

Modern man, Western or Westernized, is not going to snuggle back into the bosom of nature, perceiving all reality as a blurred continuum. That possibility of innocence we lost long ago in—of all places—a garden. We have understood differentiation, specialization, individuation; we have known the glories of action concentrated upon a specific purpose. Our path toward unity lies *through* diversity and specialization, not in recoil from them. A high-technology society without adequate institutions of coordination will produce either chaos or tyranny or both. Freedom will become meaningless because individual men will cease to believe that what they want has any relevance to what they get. But a high-technology society that can innovate adequate structures of decision will expand the freedom of individual choice far beyond any dream of the low-technology centuries.

The chief product of the future society is destined to be not food, not things, but the quality of the society itself. High on the list of what we mean by quality stands the question of how we deal with the material world, related as that is to how we deal with one another. That we have the wealth and the power to achieve a better environment is sure. That we will have the wisdom and charity to do so remains—and must always remain—uncertain.

70 71 72 73   12 11 10 9 8 7 6 5 4 3 2 1